MASS SPECTROMETRY

McGRAW-HILL
INTERNATIONAL
BOOK COMPANY

New York
St Louis
San Francisco
Auckland
Bogotá
Guatemala
Hamburg
Johannesburg
Lisbon
London
Madrid
Mexico
Montreal
New Delhi
Panama
Paris
San Juan
São Paulo
Singapore
Sydney
Tokyo
Toronto

IAN HOWE

Senior Research Assistant, University College, Swansea

DUDLEY H. WILLIAMS

Fellow of Churchill College, Cambridge

RICHARD D. BOWEN

Fellow of Sidney Sussex College, Cambridge

MASS SPECTROMETRY
Principles and Applications
Second Edition

This book was set in Times Roman Series 327

6373 - 0820

Chemistry Lib.

British Library Cataloguing in Publication Data

Howe, Ian
  Mass spectrometry.
  1. Mass spectrometry
  I. Title   II. Williams, Dudley Howard
  III. Bowen, Richard D
  543'.0873   QD96.M3   80-41864

  ISBN 0-07-070569-0

MASS SPECTROMETRY

1 2 3 4 MM 8 3 2 1

Typeset in Great Britain by Advanced Filmsetters (Glasgow) Ltd
Printed and bound in the United States of America by the Maple-Vail Book Manufacturing Group

CONTENTS

PREFACE

This book is not written for the expert in mass spectrometry. Rather, it is written for all those who may want to know how a mass spectrometer can help them in studies in chemistry, biochemistry, and medicine. Throughout, we have used a relatively simple presentation and concentrated on the principles involved. These principles are then illustrated by means of a few selected examples and applications.

We have attempted to strike a balance between theory and applications, although being aware that the majority of practitioners will be interested in structural and analytical applications. Thus, theoretical considerations (Chapters 2 to 5) have been kept to a minimum, but hopefully are sufficient to avoid a "black-box" approach by the following generation of users. We believe that the omission of theory and principles could lead to a situation where future advances of a fundamental nature may be missed or their discovery retarded.

Our sincere thanks are expressed to those authors mentioned in references, not only for making the results of their researches available through publication but also in many cases for helping us to understand the importance of their work through personal contacts. We are also indebted to the following journals and publishers for permission to reproduce figures: *Zeitschrift für Naturforschung*; *Journal of the American Chemical Society*, *Journal of Organic Chemistry*, *Journal of Physical Chemistry*, *Accounts of Chemical Research*, *Analytical Chemistry*, and *Inorganic Chemistry* (American Chemical Society); *Journal of Chemical Physics* (American Institute of Physics); *Organic Mass Spectrometry* (Heyden and Son Ltd.); *International Journal of Mass Spectrometry and Ion Physics* and *Journal of Chromatography* (Elsevier Publishing Company); *Angewandte Chemie* and *Chemische Berichte* (Verlag Chemie GmbH); *Experientia* (Birkhauser Verlag);

Journal of the Chemical Society: Chemical Communications (The Chemical Society, London); Interscience Publishers, John Wiley and Sons, Inc.; The Institute of Petroleum, London; Holden-Day, Inc.; *Canadian Journal of Chemistry, Canadian Journal of Biochemistry* (The National Research Council of Canada); *Analytical Biochemistry* (Academic Press).

We express our gratitude to Mrs. E. Bowen, Dr. H. M. Swain, and Mrs. P. Williams for their assistance in preparing the manuscript.

Ian Howe
Dudley H. Williams
Richard D. Bowen

ONE

MASS SPECTROMETERS

Mass spectrometers, in their simplest forms, are designed to perform three basic functions. These are (1) to vaporize compounds of widely varying volatility, (2) to produce ions from the neutral molecules in the gas phase, and (3) to separate ions according to their mass-to-charge ratios (m/z), detect, and record them. The devices used to perform these functions are described briefly in the following sections, dealing first with the vaporization of the sample.

1-1 INLET SYSTEMS

The introduction of a gaseous sample into a mass spectrometer usually presents no problem. All that is necessary is to connect the gas container to the ion source using a suitable gas line and evacuate the air from this inlet system. Careful manipulation of the valve on the gas container then admits enough gas into the ion source to obtain a mass spectrum. In order to minimize the possibility of admitting too much sample, a reservoir and a sinter are usually installed to control the flow of gas into the ion chamber.

A simple modification of such a cold inlet system is all that is required to permit liquid samples to be introduced into the ion source. A typical heated inlet system is depicted in Fig. 1-1.

When not in use, the inlet system is left with valves A and D closed whilst valves B and C are kept open. Hence, the reservoir and most of the glass lines connecting the ampoule and the ion source are maintained under vacuum.

In order to admit a volatile liquid sample, valves B and C are closed as a precautionary measure and the ampoule is removed. A suitable amount of sample,

To ion source

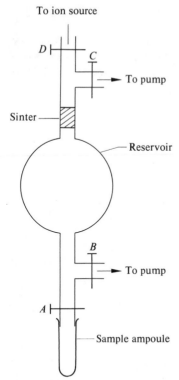

Figure 1-1

usually of the order of 1 mg, is inserted into the ampoule; it is then replaced on the inlet system and cooled in liquid nitrogen. Valves *A*, *B*, and *C* are progressively opened and the inlet system is evacuated; the reservoir is then isolated from the pumps by closing valves *B* and *C*. Valve *D* is now opened and the liquid nitrogen cooling the ampoule is removed, thus allowing the liquid to evaporate into the reservoir and then pass through the sinter into the ion source. When a sufficiently high pressure (for example, 10^{-6}–10^{-7} torr) of gas is registered by gauges in the ion source, valve *A* is closed leaving enough vaporized sample in the reservoir to obtain a mass spectrum.

Somewhat less volatile samples may be admitted by heating either the inlet system, or the sample ampoule, or more usually both. Normally, the inlet system is maintained at a higher temperature than the ampoule, which may be heated independently. Progressively increasing the temperature of the ampoule then avoids unnecessary thermal decomposition of the liquid sample. Moreover, since these heated inlet systems are frequently made entirely of glass, contact between the sample and potentially active metal surfaces is also prevented. Bearing in mind that the whole inlet system may be heated to temperatures as high as 300 or 350°C and simultaneously evacuated to 10^{-1} torr, it is clear that such a system is extremely versatile and may be used to introduce a wide variety of liquids into the ion chamber. In addition, it is possible to use the same system to admit solid samples into the ion chamber provided that these samples have a sufficiently high vapor pressure to permit volatilization.

Heated inlet systems which do not involve a sample ampoule may be constructed. One such system employs a rubber septum in place of the ampoule; a liquid sample is then introduced into the reservoir via a microsyringe, which is inserted through the septum. Precise amounts of sample may be admitted in this way. Another system uses a liquid metal to glass seal instead of the ampoule; the liquid sample is placed in a fine capillary, which is then pushed through the liquid metal, so as to touch the glass sinter underneath. The sample is then free to pass through the sinter into the reservoir, whilst air is excluded by the liquid gallium seal.

Frequently it is desirable to deal with samples that have insufficient thermal stability to be heated to 200–300°C, or that have very low vapor pressures even at these temperatures. Many substances in the molecular weight range 400–1200 daltons fall into the latter class. For these compounds, advantage is taken of the extremely low pressures, of approximately 10^{-7} torr, which may be achieved in the ion chamber. The sample is applied to the end of a probe which is then introduced directly, via a vacuum lock, into the ion chamber. The sample is often heated simply because of its proximity to the electron beam; in addition, further heating may be performed by means of an electrical element in the probe. Carefully controlled heating of the probe is sometimes necessary, so that relatively involatile samples are not ejected into cold parts of the source without coming into contact with the ion beam. Controlled heating also avoids charring of the sample which can result if the probe temperature is raised too rapidly. A vapor pressure of only about 10^{-6} torr permits a spectrum to be obtained, using the direct insertion technique, from a sample of a few micrograms or less. The minimum amount of sample required to obtain a spectrum should be used, in order to maximize the source lifetime (the time before cleaning the source is necessary). In addition, if excessive amounts of samples are employed, the source slits will become dirty, thus decreasing sensitivity and resolution. Moreover, high-voltage breakdown may occur in the source, and eventually the pump oil will be contaminated. Another precaution, which can be employed to increase the source lifetime, is to avoid generating an ion beam when it is not required. For example, when instrument adjustments or analysis of spectra are being performed whilst running a compound, it is advantageous to withdraw the probe and reduce the temperature slightly; unnecessary vaporization of the sample is thus avoided.

Relatively volatile contaminants can often be removed from the source by applying gentle baking overnight, or during other periods when the instrument is not in use. After the source has been removed for more thorough cleaning, adjustment, repair, or replacement by another source, baking at higher temperatures is normally employed.

In the case of extremely involatile samples, it is sometimes possible to obtain a spectrum by introducing the probe tip into the edge of the ionizing electron beam. This modification of the direct insertion method possibly involves ionization of some sample molecules without prior volatilization (see Sec. 6-9).

Other more sophisticated methods are now available for volatilization and ionization of extremely involatile compounds; these are discussed in Sec. 1-2. It is also possible to link gas or liquid chromatography columns directly to the source of a mass spectrometer. These powerful techniques, which require modification of the inlet system, are discussed in Secs. 8-2 and 8-3.

1-2 ION SOURCES

The region of the mass spectrometer where ions are generated is called the ion source. A variety of techniques may be applied to ionize molecules of the vaporized sample. In this section, some of the more important methods are outlined and their relative advantages and disadvantages are briefly discussed.

A. Electron Impact

The most common technique currently in use for the production of positive ions is electron impact (EI). A few milligrams of the substance to be examined are introduced, as a vapor, into the source at the operating pressure (ca. 10^{-6} torr). The vapor is allowed to pass through a slit, A, into the ionization chamber (Fig. 1-2), where it is bombarded with a beam of electrons accelerated from a hot filament. The energy of the electron beam can be varied from 0 to 100 eV. The ionization energies of most organic molecules fall in the range 7–13 eV ($1 \text{ eV} \simeq 23 \text{ kcal mol}^{-1} \simeq 96 \text{ kJ mol}^{-1}$); consequently, it is possible to supply energy in excess of the ionization energy. The ionization energy corresponds to the energy required to remove an electron from the highest occupied molecular orbital; this reaction is represented as follows:

$$M + e \rightarrow M^{\ddagger} + 2e \qquad (1\text{-}1)$$

The $M^{\ddagger}$ symbolism denotes that the molecular ion, thus formed, is both a positive ion and a radical (since organic molecules are almost without exception even-

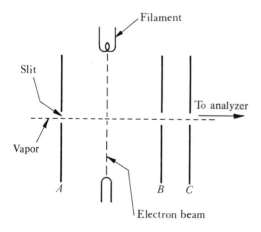

Filament

Slit

To analyzer

Vapor

A B C

Electron beam

Figure 1-2

electron species). The alternative process, electron capture by a molecule of vapor to give a negative-ion radical

$$M + e \rightarrow M^{-} \qquad (1\text{-}2)$$

is usually less probable. Negative-ion mass spectrometry has been less extensively developed as a method of structure elucidation for organic molecules; moreover, energetic and kinetic studies on negative ions are not as far advanced as the corresponding studies on positive ions. This book concentrates mainly on positive-ion mass spectrometry; negative-ion mass spectrometry is discussed in Secs. 2-2 and 7-2.

The energy given to the molecules of vapor, by the ionizing electrons, is usually greater than the ionization energy. Therefore, excess energy remains in the molecular ion (M^{+}) and this excess energy may be used to break one (or possibly more) bond(s). Depending on the energy content, molecular ions may remain as such or dissociate into fragment ions. In general, a molecular ion may dissociate by elimination of a radical

$$M^{+} \rightarrow A^{+} + B\cdot$$

or by loss of an even-electron molecule

$$M^{+} \rightarrow C^{+} + D$$

In either event, the species $B\cdot$ and D are thereafter ignored by the mass spectrometer since they do not possess a positive charge. In the former case the fragment ion is an even-electron species, while in the latter case the fragment is an odd-electron ion radical. Depending on their internal energies, the fragment ions A^{+} and C^{+} also have the possibility of fragmenting.

The extent to which the molecular ion fragments is usually relatively large when electron-impact ionization is employed. It is sometimes not possible to detect an appreciable abundance of the molecular ion. This problem, which is often encountered for molecular ions which can undergo extremely facile fragmentations, is a serious drawback of electron-impact ionization. In some cases, it is possible to increase the relative abundance of the molecular ion by reducing the energy of the ionizing beam of electrons. When the energy of the ionizing electrons approaches the ionization energy of the molecule, the excess energy remaining in the molecular ion is small and fragmentation is decreased.

However, the large degree of fragmentation, associated with electron-impact ionization, is often advantageous. In fact, a wealth of information concerning the structure of the molecule under investigation may be deduced from the fragmentation pattern. This subject is treated in more detail in Chap. 6.

B. Chemical Ionization

Chemical ionization (CI)[1] is defined to occur in the mass spectrometer when a molecular sample is ionized via an ion-molecule reaction. Ionization is thus effected by chemical reaction rather than by electron bombardment. The ions

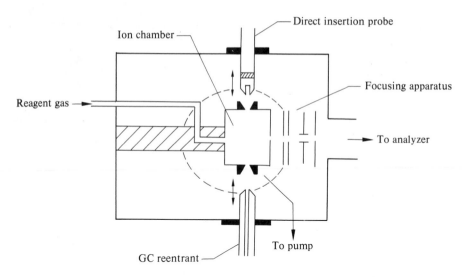

Figure 1-3

required for this process are called reagent ions and are normally generated from a reactant gas in a high-pressure ion source. These reagent ions are usually present in large excess, compared with sample molecules, in order to minimize any electron-impact ionization of the sample molecules.

A schematic diagram of a typical chemical ionization source is shown in Fig. 1-3. A reagent gas (e.g., methane, isobutane, or ammonia) is allowed to pass into the ion chamber and the sample is introduced in the normal way; the direct insertion probe, or a gas-chromatography (GC) inlet system, is usually employed for this purpose. Good chemical ionization conditions are attained when a high concentration of reagent ions is present in the ion chamber. This may be achieved by using ionizing electrons with energies up to 300 eV together with a pressure of about 1 torr in the ion chamber. On account of the high pressures involved, a good source design must incorporate a gas-tight ion chamber and sufficient pumping capacity to reduce collisions, outside the ion chamber, to a minimum.

The internal energy content of the molecular ion (or pseudomolecular ion, e.g., MH^+), produced by chemical ionization, is generally much less than is the case for electron-impact ionization. Furthermore, the even-electron MH^+ ions, formed by chemical ionization, are usually more stable toward fragmentation than are the odd-electron ions, produced by electron impact. Consequently, a smaller degree of fragmentation is often observed for chemical ionization than is found for electron-impact ionization. Therefore, there is a greater probability of detecting a molecular ion, or a species which is very closely related to the neutral molecule; hence, chemical ionization may yield molecular weight, or molecular formula, information when electron bombardment does not.

In addition, chemical ionization is a sensitive technique, even for negative ion formation in appropriate cases; this sensitivity, coupled with the "soft" nature of the ionization process, is a major advantage of the method.

C. Field Ionization

In addition to ionizing molecules by electron bombardment (Sec. 1-2A) or by chemical reactions (Sec. 1-2B), it is also possible to use powerful electric fields to promote ionization. This technique is known as field ionization (FI).[2]

The principle of the method is to allow the vaporized sample to approach within close vicinity of an intense electric field (of the order of 10^9–10^{10} V m^{-1}). This field may be achieved by using metal anodes (e.g., made of tungsten), which have very small radii of curvature. Suitable shapes for the anodes include points, blades, and fine wires. The anode may also be modified by etching, by growing whiskers on the surface, or by other chemical treatment, in order to improve the quality of the surface. A distinct advantage of using a fine wire is that the resultant ion current is often greatly increased, on account of the relatively large surface area of the anode. Frequently wires of roughly 10 μm diameter are employed which are activated by growing surface microneedles with the aid of benzonitrile.[3]

A typical field ionization source having a wire anode and a plane slotted cathode is depicted schematically in Fig. 1-4. At sufficiently high field strengths, essentially every molecule which enters the region of intense electric field undergoes field ionization. The ionization of the molecules occurs by migration of an electron to the anode. This process is based on the quantum mechanical tunneling effect.

When field ionization is used, the quantity of excess energy, imparted to the molecular ion, is usually small. Field ionization is thus a "soft" ionization technique and relatively abundant molecular ions are usually observed. Therefore, molecular weight or molecular formula information can be acquired with relative ease. However, on account of the rather low degree of fragmentation, less structural information is accessible than is the case for electron-impact ionization. Moreover, the total ion current is often relatively small; this insensitivity arises from the poor sample ionization efficiency and the problems associated with focusing the beam of the ions produced.

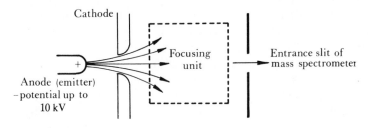

Figure 1-4

D. Field Desorption

The three ionization techniques described above involve vaporization of the sample followed by ionization of the resultant molecules in the gas phase. Consequently, when the compounds to be investigated are either extremely involatile or thermally unstable, application of these methods may fail to produce a useful mass spectrum. For such compounds, it would clearly be better to ionize the molecules prior to volatilization. This may be achieved by means of field desorption (FD),[4] which is essentially a modification of field ionization.

The sample for field desorption is deposited in a thin film on the surface of an anode (usually a fine wire) of a conventional field ionization source. The anode may be immersed in a solution of the sample or, alternatively, a drop of sample solution may be loaded onto the anode using a microsyringe. The anode is then inserted into the field ionization source through a vacuum lock and the field is switched on. Migration of an electron from a sample molecule to the anode may occur and the resulting cation is strongly repelled by the anode. The ion thus formed is therefore desorbed and may subsequently leave the ion source and enter the analyzer.

The enormous advantage of field desorption is that it is not necessary to vaporize the sample by heating it, although passage of a small current through the emitter wire is normally required. Molecular ions may be obtained by field desorption of biomolecules which could not withstand the temperatures that would be needed to vaporize these compounds in electron-impact, chemical ionization, or simple field ionization sources.

A major disadvantage of the method is that the total ion current obtained is normally very low compared to other ionization methods.

E. Plasma Desorption

The thermal degradation of heat-sensitive compounds can, in principle, be overcome by extremely rapid heating. If the temperature rise is sufficiently rapid, there is not enough time for thermal decomposition of the sample to occur before vaporization. This effect is exploited in the plasma desorption source illustrated in Fig. 1-5.

The sample to be analyzed is deposited on a thin metal foil, usually of nickel, which is placed in the ion source in precise alignment with a ^{252}Cf fission fragment source. Spontaneous fission of the ^{252}Cf nuclei can occur and each fission event gives rise to two fragments, with unequal masses and energies, traveling in almost exactly opposite directions. The small departure from colinearity is caused by neutron emission. Approximately 40 different pairs of fragments can result from spontaneous fission of ^{252}Cf. A typical pair of fission fragments is ^{142}Ba^{18+} and ^{106}Tc^{22+}; the kinetic energies associated with these fragments are roughly 79 and 104 MeV, respectively. When such a high-energy fission fragment passes through the sample foil, extremely rapid localized heating occurs, producing a temperature in the region of 10,000 K. Consequently, the molecules in this plasma zone are

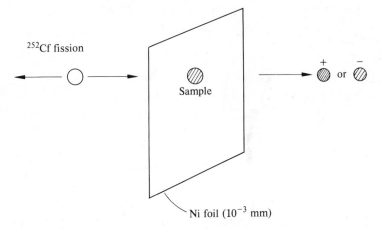

Figure 1-5

desorbed, with the production of both positive and negative ions. These ions may then be accelerated out of the source into the analyzer system. In the apparatus used to date,[5] ion analysis is achieved by means of the time-of-flight principle (Sec. 1-3C); the complementary fission fragment to that causing ionization is utilized as a zero time marker.

The apparatus required to perform plasma desorption is highly specialized and is not as yet commercially available. On the other hand, this technique is extremely useful for obtaining mass spectra of thermally sensitive or involatile compounds. It is potentially capable of yielding molecular ions (or ions which are simply related to the molecular ion, such as MH^+) for biomolecules of relatively high mass (for example, 3000 daltons; see Sec. 7-4).

1-3 MASS ANALYSIS

Once the sample to be investigated has been volatilized and ionized, it is necessary to analyze the ions which are produced. In general, the ions are repelled out of the source using a small (less than 50 V) repeller voltage, transmitted through an accelerating potential (where required) and then injected into a mass analyzer. These mass analyzers, which separate ions according to their mass-to-charge ratio, may take several forms. Some of the more important methods of mass analysis are discussed below.

A. Magnetic-Sector Instruments

Dempster's first mass spectrometer[6] was of the single-focusing type in which the positive ions were deflected through 180° in a magnetic field B (see the schematic illustration in Fig. 1-6).

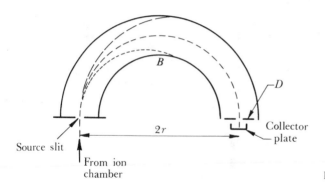

$2r$

Source slit

From ion
chamber

D

Collector
plate

Figure 1-6

As might be expected intuitively, the heaviest ions are constrained into the circular paths of largest radius. Quantitatively, any ion having a charge ze when accelerated through a voltage V will acquire a translational (kinetic) energy zeV, where e is the charge of the electron and z is the number of such charges. Thus, the kinetic energy of the ions is independent of their masses, but since

$$\tfrac{1}{2}mv^2 = zeV \tag{1-3}$$

(where m is the mass of the ion and v its velocity after acceleration), the more massive ions travel more slowly. In a magnetic field of strength B, any ion will experience a force of $Bzev$, producing an acceleration of r^2/r toward the center of a circular path of radius r. Hence, from Newton's second law of motion,

$$Bzev = \frac{mv^2}{r} \tag{1-4}$$

Combining Eqs. (1-3) and (1-4) leads to

$$\frac{m}{z} = \frac{B^2 r^2 e}{2V} \tag{1-5}$$

From Eq. (1-5) it is evident that, at a given magnetic field strength and accelerating voltage, ions of a given m/z value will follow a particular path of radius r and that the ions of various m/z values may be progressively transmitted through the magnetic field and the collector slit (Fig. 1-6) either by varying B at constant V or by varying V at constant B. The former possibility is called magnetic scanning and the latter electric (or voltage) scanning. Either mode may be used in practice.

The angle of deflection does not necessarily have to be 180° to retain the ion-beam focusing properties of magnetic fields. Indeed, sector fields are cheaper to produce and have the advantage that the ion source and collector are more accessible for maintenance than in 180° instruments. A diagram of a 90° sector single-focusing mass spectrometer is given in Fig. 1-7. The term "single focusing" indicates that the instrument focuses the ions either in direction or in velocity; the machine illustrated in Fig. 1-7 performs directional focusing.

One of the problems of the sector instrument shown in Fig. 1-7 is that the resolving power is limited by the initial spread of translational energies of the ions

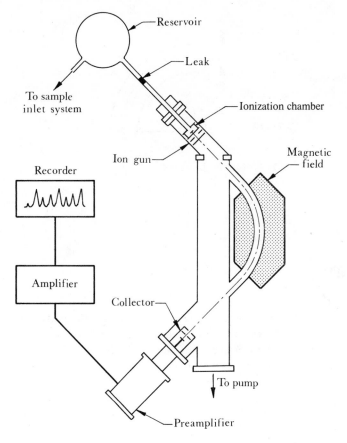

Figure 1-7

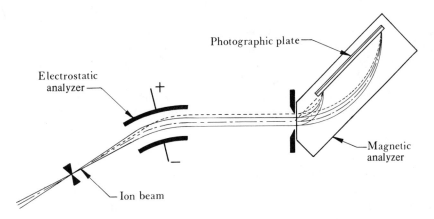

Figure 1-8

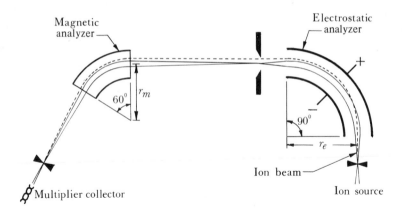

Figure 1-9

on leaving the source (due to the Boltzmann distribution and field inhomogeneity in the source). This problem may be overcome by passing the ions through an electric field prior to the magnetic field. The electric field effects velocity focusing of the ions; thus, by placing a slit between the electrostatic and magnetic analyzers, ions of a closely defined kinetic energy may be selected prior to mass analysis. Instruments incorporating such a system are described as double-focusing mass spectrometers; this is because the ions are focused twice, in direction and in energy. These mass spectrometers are capable of attaining a much higher resolving power than is possible with single-focusing instruments.

In the instrument of Mattauch and Herzog,[7] ions of all m/z ratios are brought to a double focus in a plane and a photographic plate can then be used to record the ions (Fig. 1-8).

On the other hand, in the geometry used by Johnson and Nier,[8] shown in Fig. 1-9, only electrical detection is possible, since there is only one double-focusing point. An electron multiplier collector is usually employed.

Instruments having either Mattauch-Herzog or Nier-Johnson geometry are commercially available. High-resolution studies on high-molecular-weight organic compounds may be performed routinely using such instruments.

Double-focusing mass spectrometers may also be constructed such that the ions are transmitted by the magnetic sector before entering the electric sector. Since this is the reverse order to that of the instruments depicted in Figs. 1-8 and 1-9, mass spectrometers are said to have reversed geometry.[9] There are several advantages associated with the use of this geometry, especially in elucidating the decomposition channels of a given ion. For example, the ion of interest may be transmitted exclusively through the magnetic sector and all daughter ions arising from decomposition of this ion, occurring between the two sectors, may be discovered by scanning the electric-sector voltage (Fig. 1-10). The resulting mass-analyzed ion kinetic energy spectrum (MIKES) may be used to "fingerprint" particular ion structures (for details see Sec. 3-2B). Using this technique, it is

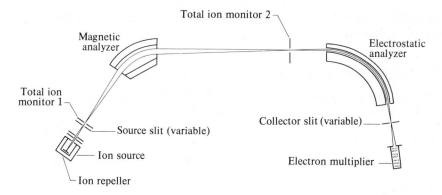

Figure 1-10

possible to find all the daughter ions formed from a given parent ion. The same information may be obtained using machines of conventional geometry; however, these instruments acquire the data in the opposite sense. The magnetic sector is tuned to allow the passage of a given ion. All dissociations in the region prior to the electric sector which give rise to this ion are then found by scanning the electric sector voltage (or accelerating voltage). This permits all parent ions which dissociate to form a given daughter ion to be documented. It is important to grasp that the same total information is available from a series of such high-voltage (HV) scans, on suitable daughter ions, as is accessible from MIKE spectra of the corresponding parent ions. The MIKES technique is often considered more convenient in practice.

It is also possible to scan the magnetic and electric sectors simultaneously; this procedure is known as a linked scan. Several different linked scans are possible,[10,11] the most useful being that in which the ratio of the magnetic and electric fields is kept constant.[10] When such a B/E linked scan is performed using a conventional or reversed geometry instrument, the spectrum reveals all daughter ions formed by decomposition of the selected parent ion. However, the linked scan does not retain any information concerning the quantity of kinetic energy which is released upon fragmentation (see Sec. 3-2D). In this respect, it differs from the MIKES technique which does yield kinetic energy release data. This feature is a potential disadvantage for certain types of investigations, especially ion structure elucidation; on the other hand, it is advantageous from the analytical point of view, because sharp, nonoverlapping peaks are observed.

The analysis of negative ions by magnetic-sector instruments follows the same principles as those for positive ions. The negative ions are usually produced in a chemical ionization source (Secs. 1-2B and 2-2D). The polarities of the accelerating and focusing voltages must be reversed, relative to those employed in positive-ion operation. In a double-focusing mass spectrometer, the direction of the field between the electric-sector plates must also be reversed. The field in the magnetic sector is reversed by reversing the current direction in the magnet coils. At the

highest levels of convenience, all these reversals can be accomplished by manipulation of a single switch.

B. Quadrupole Mass Spectrometers

There are several alternative methods available for mass separation other than
deflection in a magnetic field as described above. Mass spectrometers currently in
use which employ other separation methods include the time-of-flight mass
spectrometer (Sec. 1-3C) and the ion-cyclotron resonance instrument (Sec. 1-3D).
However, the technique which has provided the most successful commercial
competition with magnetic analysis in recent years is the quadrupole mass filter.

Quadrupole mass spectrometry evolved from techniques used in particle
acceleration. It was found that proton beams could be focused by using alternating
quadrupole fields and later the recognition that focusing could be made to be mass
selective led to the development of the quadrupole mass filter.[12]

Figure 1-11 depicts the arrangement of electrodes required in the quadrupole
mass filter. A constant dc voltage U and a radiofrequency potential $V \cos \omega t$ are
applied between opposite pairs of four parallel rods of hyperbolic cross section
(usually of ceramic construction). The rods are in most cases between 0.1 and 0.3 m
long in commercial instruments.

Ions are injected from the ion source into the quadrupole analyzer along the x
direction and are acted upon by a variable field in the yz plane. The equations of
motion for an ion in the quadrupole analyzer are considerably more complex than
in the magnetic analyzer described in Sec. 1-3A. It emerges from the solution of

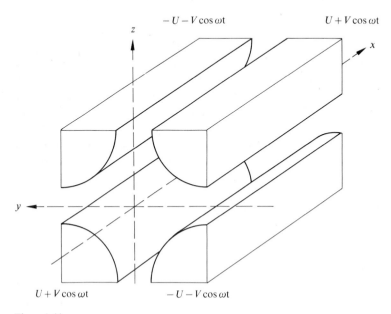

Figure 1-11

these equations that for a given set of voltages (U,V), frequency ω and rod separation, a stable condition may exist such that an ion trajectory remains within the bounds of the electrodes. Such ions emerge at the exit from the quadrupole mass filter to be detected in the usual way. This stable condition applies to ions having a range of m/z values and obviously in practice this range is made sufficiently narrow to transmit only ions of the same nominal m/z value. Ions of other m/z ratios undergo unstable oscillations and are removed in the y and z directions, either by impinging on the electrodes or by escaping between the rods. The mass spectrum is scanned and ions of other m/z ratios brought into the condition of stability either by varying the amplitude of V and U, while keeping the ratio V/U constant, or by varying the frequency ω of the rf potential (which is usually several megahertz).

Correct spacing of the rods (to within a few microns) is essential for mass resolution and efficient transmission of ions. Rods of circular cross section have frequently been employed because of difficulties in fabrication of hyperbolic rods with high precision; however, an increasing number of quadrupole mass spectrometers with hyperbolic rods are now becoming commercially available.

The mass range of a quadrupole mass spectrometer does not normally extend above m/z 1000, although commercial instruments have recently become available which perform satisfactorily above this mass. A problem arises in achieving the required sensitivity at high mass. If scan parameters are adjusted to achieve constant peak width and separation throughout the mass range, heavier ions are transmitted less efficiently and sensitivity loss at high mass results. The quadrupole mass spectrum normally maintains unit resolution throughout the mass range, that is, 100 at m/z 100, 1000 at m/z 1000. A resolution of several thousand is now achievable in commercial instruments, especially where hyperbolic rods are employed. Mass resolution is discussed in Sec. 1-4.

Quadrupole instruments are commercially successful because they are relatively inexpensive, compact, and require little experience to operate and maintain. Where sensitive, low-mass analysis is required, instruments are available to suit the particular mass range (e.g., quadrupole gas analyzers are available which have an upper mass limit of m/z 60). Another advantage is the fast scanning capability (up to 1000 daltons s^{-1}), which lends itself to GC/MS applications (see Sec. 8-2). The linear mass scale of the quadrupole is also an advantage in data handling and in computer control of the mass spectrometer scan. Furthermore, the absence of high voltages facilitates smooth operation.

It is only in the field of applications which require high-resolution or high-mass sensitivity that the magnetic analyzer becomes essential. Quadrupole design is constantly being improved, but it is unlikely that quadrupoles will replace magnetic instruments for high-resolution work in the foreseeable future.

A quadrupole mass filter is capable of the simultaneous analysis of both positive and negative ions. Indeed, the simultaneous production of both positive and negative ion mass spectra can be achieved using a pulsed ion source.[13] The ion source potential (± 4–10 V) and focusing lens potential (± 10 V) are pulsed at a rate of 10 kHz (see Fig. 1-12). Under these conditions, packets of positive and

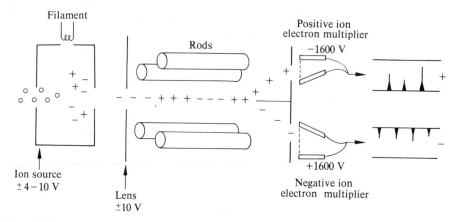

Figure 1-12

negative ions are ejected from the source in rapid succession and enter the quadrupole. Detection of the positive and negative ion beams is accomplished by means of two electron multipliers, situated side by side at the exit aperture of the quadrupole rods. By applying potentials of opposite sign on the two multipliers, positive ions are attracted to one multiplier and negative ions to the other. The result is that positive and negative ions are recorded simultaneously as deflections in opposite directions by an ultraviolet light detector (Sec. 1-4).

C. Time-of-Flight Mass Spectrometers

The time-of-flight mass spectrometer[14] (see Fig. 1-13 for a schematic diagram) employs as its method of mass separation the principle that ions of different mass, accelerated to a uniform kinetic energy, have different velocities, and hence different times of flight over a given distance. The most common type of time-of-flight spectrometer uses a pulsed ion source with a voltage pulse on grid A to extract the ions from the source. A potential gradient between A and B accelerates the ions to uniform kinetic energy and into the field-free flight tube. The ions are

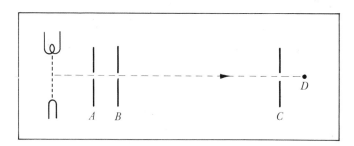

Figure 1-13

separated in time, according to their m/z values, on arrival at the repeller grid C and are collected at D. It is common to have differences in arrival times between successive mass peaks of 10^{-7} s, or less; consequently, fast electronics are required to distinguish successive peaks.

D. Ion-Cyclotron Resonance Spectrometers

Ion-cyclotron resonance (ICR) mass spectrometry[15,16] is a powerful technique for investigating ion-molecule reactions; it is especially useful for elucidating the structures of ions in the gas phase. ICR may be used to observe collision processes on account of the long ion lifetimes (typically milliseconds, though even longer lifetimes are feasible) which are involved. In contrast, collisions are achieved in chemical ionization mass spectrometry (Sec. 1-2B) by the use of relatively high source pressures.

A schematic view of a typical ICR cell is shown in Fig. 1-14. Ions are generated by electron impact in a uniform magnetic field B (perpendicular to the plane of the paper) and are restricted to a circular path perpendicular to the direction of the magnetic field. The angular frequency ω_c of this motion is independent of the velocity of the ions and is given by

$$\omega_c = zeB/m \tag{1-6}$$

where m/z, as usual, is the mass-to-charge ratio for a given ion. The angular frequency ω_c therefore depends on the m/z value and is known as the cyclotron resonance frequency. If an alternating electric field of radiofrequency ω_1 is applied normal to B, an ion will absorb energy if $\omega_c = \omega_1$, and an ICR mass spectrum can

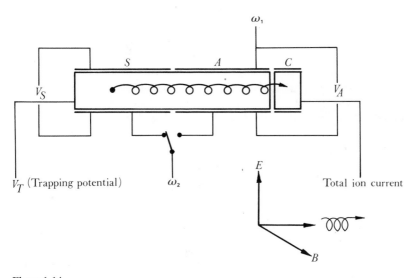

Figure 1-14

be obtained by fixing ω_1 and scanning the magnetic field B so that ions of different m/z satisfy Eq. (1-6). Absorption of energy by the ions (at resonance) is measured using an oscillator-detector system similar to those used in NMR and ESR spectroscopy.

After formation in the source S, ions are made to pass through the analyzer A to the collector C by application of a small static electric field E, perpendicular to B. The velocity of the ions, in a direction perpendicular to both B and E (see Fig. 1-14) is given by

$$v = \frac{E}{B} \tag{1-7}$$

In the final region of the ICR cell, the total ion current is measured at the collector. Some typical parameters for an ICR mass spectrometer are given below:[16]

Operating pressure $= 10^{-7}$–10^{-4} torr
Length of source region $=$ length of analyzer region $= 6.35 \times 10^{-2}$ m.
For Ar^+, m/z 40, and for $B = 0.8$ tesla and $E = 25$ V m^{-1}:
$$\omega_c = 307\,\text{kHz}$$
$$v = 30\,\text{m s}^{-1}$$
Time spent in analyzer region $= 2 \times 10^{-3}$ s

Since the analyzer residence times are of the order of milliseconds, there is a high probability of observing ion-molecule reactions even at low pressures. Moreover, the application of various double-resonance and pulsing techniques facilitates the elucidation of these reactions.

The double-resonance technique[17] is very helpful because it may be used to deduce the precursor of a given daughter ion. Consider the general ion-molecule reaction, $A^+ + B \rightarrow C^+ + D$. The magnetic field is tuned so that the C^+ ions absorb energy at an oscillator frequency of ω_1. A second radiofrequency field of frequency ω_2 is then applied to the source of the ICR cell and ω_2 is scanned until the cyclotron resonance frequency of A^+ is attained. At this frequency, the A^+ ions absorb energy and consequently their translational energy increases. The rate constants of ion-molecule reactions involving A^+ depend on the translational energy of the ions concerned. Therefore, the yield of C^+ should change if it is formed by ion-molecule reactions of A^+. At a fixed magnetic field,

$$\frac{\omega_1}{\omega_2} = \frac{m_A}{m_C} \tag{1-8}$$

Hence, different precursors of C^+ can be identified by scanning ω_2.

Ion-molecule reactions involving either positive or negative ions may be studied by ICR. Since the method may be used to investigate equilibrium processes, it may be applied to establish scales of basicity or acidity for numerous species. The results of such studies frequently furnish thermochemical data, e.g., the heat of reaction for $A^+ + B \rightarrow C^+ + D$. These data may be of great value in obtaining accurate heats of formation for ions in the gas phase.[18]

The use of ICR in studying bimolecular ion-molecule reactions is further discussed in Chap. 5.

1-4 RECORDING AND PRESENTATION OF MASS SPECTRA. RESOLVING POWER

After mass analysis has been accomplished, the ions arrive at the collector where they are detected, measured, and recorded. Two main methods for recording the ions are usually employed: electrical and photographic.

The term *mass spectrometer* is usually reserved for instruments in which the ion beams are measured electrically after mass analysis, while the term *mass spectrograph* refers to instruments in which the focused ion beams are recorded on a photographic plate. Mass spectrographs usually employ double-focusing methods.

The most usual presentation of a low-resolution mass spectrum (i.e., a mass spectrum in which only the integral m/z values have been resolved) is in the form of a bar graph, in which the m/z values are plotted horizontally and the ion abundances vertically. Such a bar graph for the 70 eV mass spectrum of *n*-decane (**1**) is reproduced in Fig. 1-15. The most abundant ion is arbitrarily assigned an abundance of 100 percent and is called the base peak.

If electrical recording is used, then a common arrangement is to feed the

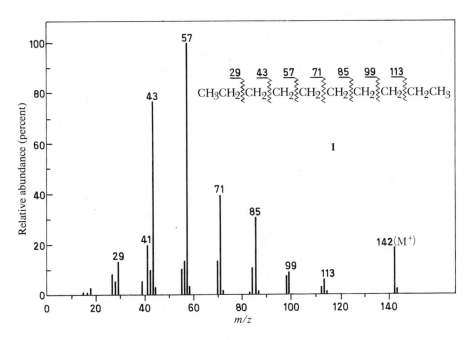

Figure 1-15

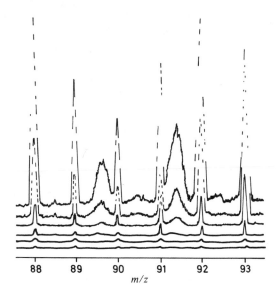

88   89   90   91   92   93

m/z

Figure 1-16

amplifier output into a recorder containing several mirror galvanometers of varying sensitivities. A beam of ultraviolet light may be reflected onto ultraviolet-sensitive paper from the mirror galvanometers. Thus, when the chart paper is driven by a motor at a constant rate during the scan of the mass spectrum, records of the mass spectrum corresponding to a wide range of sensitivities are traced out. Such a trace is given in Fig. 1-16;[19] the sensitivities of the galvanometers increase by a factor of approximately three on passing from any one trace to the one above.

If a high-resolution spectrum is recorded on a photographic plate, then the exact position and density of the lines produced by the ions on the plate are measured by means of a microdensitometer.

A recording device sensitive to very different ion-abundance ratios is essential in mass spectrometry, since ions of very low abundance will occasionally be of great importance in deducing a molecular formula or structure.

It is obvious from the large separation between the peaks in Fig. 1-16 (which is a typical *low-resolution* trace of the type which may be produced by commercially available single-focusing instruments) that the separation of m/z values differing by one unit poses no problem at m/z values of a few hundred or less. In general, a fundamental requirement of the mass spectrometer is that it resolve an ion beam of mass m from an ion beam of nearly the same mass, $m + \Delta m$. There is no single accepted definition for the situation in which m may be considered to be first resolved from $m + \Delta m$, but one method defines the resolving power as the value of $m/\Delta m$ at which the height of the valley between m and $m + \Delta m$ is 10 percent of the peak height of m and $m + \Delta m$ (Fig. 1-17). This is the so-called 10 percent valley definition. On the basis of this definition, single-focusing magnetic instruments can attain resolving powers in the region of 300–3000, or in favorable cases as high as 9000. Double-focusing instruments give resolving powers from about 10,000 to

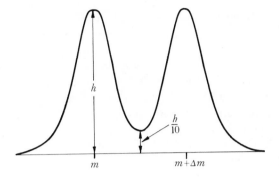

h

$\frac{h}{10}$

m $m + \Delta m$ **Figure 1-17**

40,000, although recent technical advances have pushed the latter figure up beyond 100,000. Since other definitions of resolving power (e.g., a 5 percent valley definition) are also in use, it is important to be aware of the definition employed.

The most important additions to the collection methods employed in mass spectrometry have been the computerized data-collection systems, which most commonly print out the masses and relative abundances of the ions in the spectrum. The use of computers in mass spectrometry, both for data collection and for spectral interpretation, is discussed in Chap. 9.

1-5 MASS MARKING AND PRECISE MASS MEASUREMENT

Electron-impact spectra can often be counted manually to ca. m/z 1000 without difficulty. Above this m/z value, and sometimes below, difficulties are often encountered. These problems may be exacerbated in CI spectra (Secs. 7-1 and 7-2), and it is nearly always impossible to count an FI or FD mass spectrum (Sec. 7-3). In such cases, mass marking of the spectrum is necessary. The calibration for mass marking can be done in two distinct ways: either (1) using peaks of known mass in a prior scan or (2) using peaks of known mass in the same scan. The former method may be used to calibrate a mass marker; in a magnetic sector instrument, the magnetic field is sampled by means of a Hall probe when peaks of known mass arrive at the collector. In subsequent scans, sampling of the magnetic fields which cause hitherto unknown masses to arrive at the collector is used to provide a calibrated mass scale. This scale is usually accurate to ± 1 dalton up to m/z 1000.

When peaks of known mass are generated in the same magnet scan as is used to examine the sample, the reference substance providing the peaks of known mass may either form the same ion beam or a parallel ion beam. When generating the same ion beam, sample and reference compounds are ionized in the same source. When generating a parallel ion beam, sample and reference compounds are ionized in adjacent ion sources and recorded by adjacent collectors; this technique is described in Sec. 7-3 dealing with field desorption spectra. It has an advantage that reference peaks and "unknown" peaks of the same nominal mass do not overlap. Since they are recorded in separate traces, the centroids of the peaks may be accurately determined; the relatively precise calibration scale provided by the reference beam therefore permits relatively precise mass measurement of unknown peaks, even at low resolving power.

When a single magnet scan is used to analyze ions of the sample and reference compounds, the masses of ions produced by the sample can, if necessary, be determined with even greater accuracy and atomic compositions of ions determined by linking the mass spectrometer with a computer system (Sec. 9-2F). The substances which are commonly used for mass marking spectra and accurate mass measurement are given in Tables 1-1 to 1-4, together with the main ions produced in their mass spectra.

The data of Table 1-1 are reproduced from the PFK reference spectrum. It is

Table 1-1 Partial mass spectrum of perfluorokerosene (PFK)

Mass	Peak height, %	Formula	Mass	Peak height, %	Formula
69	100.0	CF_3	481	1.8	$C_{10}F_{19}$
100	6.6	C_2F_4	493	1.5	$C_{11}F_{19}$
119	28.1	C_2F_5	505	1.3	$C_{12}F_{19}$
131	33.3	C_3F_5	517	0.8	$C_{13}F_{19}$
143	2.4	C_4F_5	531	1.2	$C_{11}F_{21}$
150	1.7	C_3F_6	543	1.1	$C_{12}F_{21}$
155	1.6	C_5F_5	555	1.1	$C_{13}F_{21}$
162	4.8	C_4F_6	567	1.0	$C_{14}F_{21}$
169	21.5	C_3F_7	581	1.0	$C_{12}F_{23}$
181	22.0	C_4F_7	593	1.1	$C_{13}F_{23}$
193	5.7	C_5F_7	605	1.1	$C_{14}F_{23}$
205	3.3	C_6F_7	617	0.9	$C_{15}F_{23}$
219	10.9	C_4F_9	631	0.7	$C_{13}F_{25}$
231	11.5	C_5F_9	643	0.7	$C_{14}F_{25}$
243	8.2	C_6F_9	655	0.9	$C_{15}F_{25}$
255	3.6	C_7F_9	667	0.8	$C_{16}F_{25}$
269	5.7	C_5F_{11}	681	0.6	$C_{14}F_{27}$
281	8.5	C_6F_{11}	693	0.6	$C_{15}F_{27}$
293	5.9	C_7F_{11}	705	0.7	$C_{16}F_{27}$
305	2.3	C_8F_{11}	717	0.8	$C_{17}F_{27}$
319	2.8	C_6F_{13}	731	0.5	$C_{15}F_{29}$
331	5.7	C_7F_{13}	743	0.5	$C_{16}F_{29}$
343	4.3	C_8F_{13}	755	0.6	$C_{17}F_{29}$
355	1.8	C_9F_{13}	767	0.6	$C_{18}F_{29}$
367	1.2	$C_{10}F_{13}$	781	0.4	$C_{16}F_{31}$
381	4.5	C_8F_{15}	793	0.4	$C_{17}F_{31}$
393	2.7	C_9F_{15}	805	0.5	$C_{18}F_{31}$
405	2.0	$C_{10}F_{15}$	817	0.5	$C_{19}F_{31}$
417	1.1	$C_{11}F_{15}$	831	0.3	$C_{17}F_{33}$
431	3.8	C_9F_{17}	843	0.3	$C_{18}F_{33}$
443	1.7	$C_{10}F_{17}$	855	0.4	$C_{19}F_{33}$
455	1.6	$C_{11}F_{17}$	867	0.4	$C_{20}F_{33}$
467	0.7	$C_{12}F_{17}$	881	0.2	$C_{18}F_{35}$

noteworthy that from $C_3F_7{}^+$ upwards (m/z 169), the ions occur in groups of *four* ions of composition C_nF_{2n+1}, C_nF_{2n-1}, C_nF_{2n-3}, and C_nF_{2n-5}, where n increases by three (in steps of one) in passing from the first to the fourth ion; for example, C_3F_7, C_4F_7, C_5F_7, C_6F_7. As a consequence, the mass differences observed in the

Table 1-2 Partial mass spectrum of perfluorotriheptyltriazine (2)

Mass	Peak height, %	Formula	Mass	Peak height, %	Formula
69	79.6	CF_3	376	5.1	$C_8F_{14}N$
76	13.8	C_2F_2N	471	4.0	$C_{10}F_{17}N_2$
100	5.3	C_2F_4	566	1.4	$C_{12}F_{20}N_3$
102	1.5	$C_3F_2N_2$	571	1.4	$C_{12}F_{21}N_2$
119	24.7	C_2F_5	616	1.2	$C_{13}F_{22}N_3$
126	9.4	C_3F_4N	771	4.8	$C_{16}F_{29}N_2$
131	19.1	C_3F_5	866	100.0	$C_{18}F_{32}N_3$
138	7.0	C_4F_4N	916	7.9	$C_{19}F_{34}N_3$
169	15.0	C_3F_7	928	9.4	$C_{20}F_{34}N_3$
176	1.3	C_4F_6N	966	2.4	$C_{20}F_{36}N_3$
181	2.2	C_4F_7	1066	1.1	$C_{22}F_{40}N_3$
219	1.3	C_4F_9	1128	1.8	$C_{24}F_{42}N_3$
269	1.4	C_5F_{11}	1166	35.5	$C_{24}F_{44}N_3$
281	1.2	C_6F_{11}	1185	2.2	$C_{24}F_{45}N_3$

Table 1-3 High mass ions in the mass spectrum of perfluorotrinonyltriazine (3)

Mass	Peak height, %	Formula	Mass	Peak height, %	Formula
971	31.3	$C_{20}F_{37}N_2$	1266	3.4	$C_{26}F_{48}N_3$
1066	100.0	$C_{22}F_{40}N_3$	1316	2.4	$C_{27}F_{50}N_3$
1166	15.4	$C_{23}F_{42}N_3$	1366	3.1	$C_{28}F_{52}N_3$
1128	28.8	$C_{24}F_{42}N_3$	1466	29.5	$C_{30}F_{56}N_3$
1166	5.8	$C_{24}F_{44}N_3$	1485	0.7	$C_{30}F_{57}N_3$
1216	2.1	$C_{25}F_{46}N_3$			

Table 1-4 Partial mass spectrum of heptacosafluorotributylamine (4)

Mass	Peak height, %	Formula	Mass	Peak height, %	Formula
219	62.0	C_4F_9	414	5.1	$C_8F_{16}N$
226	0.6	C_5F_8N	426	2.5	$C_9F_{16}N$
231	0.9	C_5F_9	464	3.8	$C_9F_{18}N$
264	10.0	$C_5F_{10}N$	502	8.6	$C_9F_{20}N$
314	0.4	$C_6F_{12}N$	538	0.4	$C_{12}F_{20}N$
326	0.4	$C_7F_{12}N$	576	1.7	$C_{12}F_{22}N$
376	0.9	$C_8F_{14}N$	614	2.6	$C_{12}F_{24}N$

PFK spectrum from m/z 169 up to m/z 319 correspond to the series 12, 12, 12, 14, 12, 12, 12, 14,... There is then a unique group of *five* ions of composition C_6F_{13}, C_7F_{13}, C_8F_{13}, C_9F_{13}, $C_{10}F_{13}$ (m/z 319, 331, 343, 355, 367). Thereafter, the mass differences again correspond to the series 12, 12, 12, 14,..., but for groups of four ions of composition C_nF_{2n-1}, C_nF_{2n-3}, C_nF_{2n-5}, C_nF_{2n-7}.

Perfluorinated compounds are used because, even when they have high molecular weights, they are relatively volatile. Furthermore, fluorine is mass deficient ($^{19}F = 18.9984022$); the peaks listed in the tables are therefore usually sufficiently below integral mass values so as to be readily resolved from C-, H-, O-, and N-containing ions of the same integral mass, even at low resolving power. Perfluorokerosene and **4** are useful for accurate mass measurement and mass marking up to ca. m/z 620 (the former somewhat higher), whereas the triazines are employed in the range m/z 800–1500 (see also Secs. 9-2B and 9-2F).

On occasions, a single mass measurement may be required with high accuracy (say to within 1 part per million). On a double-focusing instrument, this is often accomplished by the method of peak matching. The reference peak m_0 is repeatedly displayed on an oscilloscope at a given value of the magnetic field; the accelerating voltage is V_0. The peak of higher mass m_1 which is to be accurately measured will only pass through the collector slit (at the same value of the magnetic field) if the accelerating voltage V_0 is reduced to V_1 such that $m_0V_0 = m_1V_1$ [cf. Eq. (3-4)]. An additional requirement is that the initial electric sector voltage E_0 be reduced to E_1 so that ions of energy zeV_1 will be passed by the electric sector; this is achieved by coupling V and E such that a change in V is automatically accompanied by the appropriate change in E. Since $m_0V_0 = m_1V_1$, a precise measurement of the voltage ratio V_0/V_1 gives a precise measurement of the mass ratio m_1/m_0. Accurate masses of useful ions m_0 listed in Tables 1-2 to 1-4 may be calculated from C = 12.0000000, F = 18.9984022, N = 14.0030738.

Relatively precise mass measurement of ions (10–20 ppm accuracy) is possible even at low resolving power (ca. 1500) if a double-beam instrument is employed, or even with a single beam if the scan law can be reliably maintained under computer control (Sec. 9-2F). However, when compounds contain large numbers of hetero-atoms, the number of possible combinations that fall within the tolerance of the mass measurement can become unacceptably high. The problem is illustrated by the mass spectrum of a compound $C_{25}H_{54}N_5O_7Si_5P$. For a peak occurring at mass 315 in the spectrum, consider mass measurements with error limits of 1, 3, 10, and 15 ppm; the number of possible compositions of this ion (calculated from linear combinations of the atoms it may contain) are, respectively, 3, 19, 31, and 45 on the basis of these data. Therefore, structurally informative mass measurements require a mass measurement accuracy of ca. 1 ppm. Such accuracy is achieved at high resolving power (e.g., a dynamic resolving power of 30,000) employing a scan rate of 100 s per decade in mass. This is achieved by reducing the widths of the source and collector slits of a magnetic instrument. Inevitably this operation reduces sensitivity.

Details of precise mass measurement by peak matching have been considered and use of the computer to obtain complete high-resolution scans is covered in Sec. 9-2F. Precise mass measurements can also be made directly from an ultraviolet chart. A narrow scan of the desired portion of the mass spectrum is recorded by scanning the electrostatic analyzer voltage (E) of a double-focusing mass spectrometer of conventional geometry; the accelerating voltage V is linked to follow E, the scan being carried out at constant magnetic field strength. The narrow scan is advantageous if an extremely high resolution ($\geqslant 60,000$) is required over a range of a few mass units. The scan may be calibrated by the use of known reference peaks or by using the instrumentation available for peak matching (see above). In the latter technique, an accurately known voltage step $V_0 \to V_1$ allows a particular peak to be recorded twice with an accurately known mass separation (for example, 50 ppm) between the two recordings. Mass measurements may then be made directly from the ultraviolet chart.

In the mass spectra of many compounds with relatively few kinds of atoms present, a resolving power of ca. 10,000 is adequate for doublets which may arise due to mass differences $C-H_{12}$, CH_4-O, C_2H_8-S. However, if elemental compositions are required for high mass ions containing many heteroatoms, or resolution of the doublets H_2-D (mass difference 0.00155 daltons) and C_3-SH_4 (mass difference 0.0034 daltons), then resolving powers (10 percent valley definition) of 70,000 to 140,000 are usually necessary.

REFERENCES

1. M. S. B. Munson and F. H. Field: *J. Am. Chem. Soc.*, vol. 88, p. 2621, 1966.
2. H. D. Beckey: *Angew Chem. Int. Edn. Eng.*, vol. 9, p. 623, 1969.
3. H. D. Beckey, E. Hilt, A. Mass, M. D. Migahead, and E. Ochterbeck: *Int. J. Mass Spec. Ion. Phys.*, vol. 3, p. 161, 1969.
4. H. D. Beckey: *Angew. Chem. Int. Edn. Eng.*, vol. 14, p. 403, 1975.
5. R. D. Macfarlane and D. F. Torgenson: *Science*, vol. 191, p. 920, 1976.
6. A. J. Dempster: *Phys. Rev.*, vol. 11, p. 316, 1918.
7. J. Mattauch and R. F. K. Herzog: *Z. Physik*, vol. 89, p. 786, 1934.
8. E. G. Johnson and A. O. Nier: *Phys. Rev.*, vol. 91, p. 10, 1953.
9. R. P. Morgan, J. H. Beynon, R. H. Bateman, and B. N. Green: *Int. J. Mass Spec. Ion Phys.*, vol. 28, p. 171, 1978.
10. A. W. Weston, K. R. Jennings, S. Evans, and R. M. Elliott: *Int. J. Mass Spec. Ion Phys.*, vol. 20, p. 317, 1976.
11. D. L. Kemp, R. G. Cooks, and J. H. Beynon: *Int. J. Mass Spec. Ion Phys.*, vol. 21, p. 93, 1976.
12. D. Price (ed.): "Dynamic Mass Spectrometry," vol. 2, Heyden and Son, 1971.

13. D. F. Hunt, G. C. Stafford, Jr., F. W. Crow, and J. W. Russell: *Anal. Chem.*, vol. 48, p. 2098, 1976.
14. D. Price and J. E. Williams (eds.): "Time-of-Flight Mass Spectrometry," Pergamon Press, Oxford, 1969.
15. J. D. Baldeschwieler: *Science*, vol. 159, p. 263, 1968.
16. G. C. Goode, R. M. O'Malley, A. J. Ferrer-Correia, and K. R. Jennings: *Chem. in Britain*, vol. 7, p. 12, 1971.
17. J. L. Beauchamp, L. R. Anders, and J. D. Baldeschwieler: *J. Am. Chem. Soc.*, vol. 89. p. 4569. 1967.
18. J. L. Wolf, R. H. Staley, I. Koppel, M. Taagepera, R. T. McIver, Jr., J. L. Beauchamp, and R. W. Taft: *J. Am. Chem. Soc.*, vol. 99, p. 5417, 1977.
19. J. H. Beynon, B. E. Job, and A. E. Williams: *Zeitschrift für Naturforschg.*, vol. 20a, p. 883, 1965.

IONIZATION AND ENERGY TRANSFER

An important parameter associated with the various ion sources described in Chap. 1 is the transfer of energy during ionization. In particular, the amount of fragmentation of the molecular ion varies considerably according to the quantity of excess internal energy of the ion. When this excess energy is relatively large, i.e., when so-called "hard" ionization methods are employed, extensive fragmentation is frequently observed. On the other hand, when "soft" ionization techniques are used, the excess internal energy in the molecular ion is reduced and much less fragmentation usually occurs. Therefore, it is appropriate, at this stage, to examine more closely the mechanisms of the ionization methods described in the previous chapter.

2-1 ELECTRON IMPACT

A. Ionization and Appearance Energies

In considering the ionization of a molecule by electron bombardment, an important factor is the time taken for ionization compared to the time scale of molecular vibrations. An electron of energy 50 eV has a velocity of $4.2 \times 10^6 \, \text{m s}^{-1}$ and will traverse molecular diameters of a few Å (1 Å = 0.1 nm) in roughly 10^{-16} s.[1] The fastest vibrations found in organic molecules (C—H bond stretching vibrations) are almost 100 times slower. Therefore, the relative positions of the nuclei in the molecule undergoing ionization will not change during the time when the bombarding electron is in the vicinity of the molecule. Hence, when the removal of a valence electron to give a molecular ion occurs under electron impact, the process takes place without changes in the internuclear

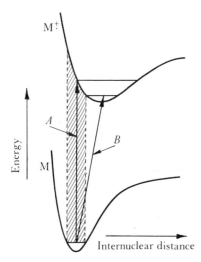

Internuclear distance   **Figure 2-1**

distances. Thus, the ionization obeys the Franck–Condon rule, which requires that the configuration and momenta of the nuclei do not alter during electronic transitions.

A simple illustration of the principle is afforded by the potential energy diagrams applicable to diatomic molecules (Fig. 2-1); analogous arguments may be applied to polyatomic molecules. Ionization in accordance with the Franck–Condon rule corresponds to the vertical transition A. The process needing the least energy corresponds to removal of an electron from the highest occupied molecular orbital. In this case, A corresponds to the vertical ionization energy, which in turn is the energy lost by the bombarding electron.

In contrast, the adiabatic ionization energy is the energy needed to remove an electron from a molecule in its lowest vibrational level to form the corresponding ion in its lowest vibrational level. This is the minimum energy required to ionize a molecule in its ground state and is represented by transition B in Fig. 2-1. The probability of obtaining such an adiabatic transition is usually very small. Consequently, electron-impact ionization energies may contain a small contribution from excess vibrational energy, which arises because the ion is not formed in its ground state. However, spectroscopic ionization energies are deduced by extrapolation of vibrational bands, which correspond to transitions to the ground state of the ion. Therefore, spectroscopic ionization energies are frequently lower than electron-impact ionization energies because no excess vibrational energy is involved. Conversely, when electron-impact ionization energies are used to determine heats of formation for gaseous ions, the values obtained may be somewhat too high on account of the excess vibrational energy present in the ion.

In addition to measuring the ionization energy (IE) of the molecular ion, it is also possible to determine the minimum energy which must be given to the molecule in order to promote the formation of a given daughter ion, D^+. This daughter ion can only appear in the mass spectrum at electron energies greater than this threshold value; consequently, such a measurement is called an appearance energy (AE). In order to evaluate heats of formation (ΔH_f) of ions from appearance energies, the structure and heat of formation must be known for each neutral (N), which is eliminated in producing the ion:

$$M + e \rightarrow M^{\ddagger} + 2e \rightarrow D^{\ddagger} + N + 2e \tag{2-1}$$

There is also the possibility that the ions or neutrals may be formed with excess energy. These problems are potentially very troublesome; nevertheless, reliable results can be obtained if suitable precautions are taken.

In practice, ionization and appearance energies may be measured, using commercial mass spectrometers, in a variety of ways. The basis of these methods, which is essentially the same in each case, involves the determination of the ionization efficiency curve for the ion in question. The ionization efficiency curve for a given ion is obtained by measuring the variation of the ion current over a range of electron beam energies. Usually measurements are begun at a nominal electron beam energy of 50 eV and the energy of the ionizing beam of electrons is

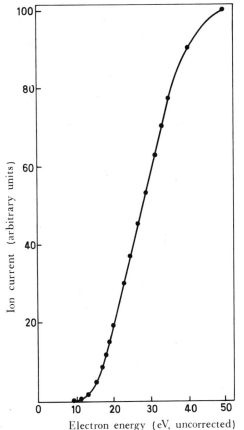

Figure 2-2

then reduced in regular increments until the ion current becomes undetectable. The ionization efficiency curve, for the production of the ion of interest, is then obtained by plotting the measured ion current against the nominal electron beam energy. A typical ionization efficiency curve is given in Fig. 2-2.

The region where the ion current falls off gradually to zero (usually within about 4 eV of the lowest energy at which measurements are possible) is described as the foot of the ionization efficiency curve. This is followed by an approximately linear rise, after which the ion current increases more slowly with increasing beam energy. The ion current is relatively insensitive to the beam energy in the range 70–150 eV.

The ionization or appearance energy of interest cannot be computed directly from the ionization efficiency curve for several reasons. One of the most important of these is that the absolute value of the electron beam energy is not known. This difficulty may be overcome by performing the same series of measurements on one or more reference substances of accurately known ionization energy. The ionization efficiency curves for these reference compounds may then be used to calibrate the electron beam energy scale; hence, the absolute value for the ionization or appearance energy may be deduced. A variety of reference compounds may be used. In general, these substances should not only have accurately known ionization energies, preferably close to the ionization or appearance energy being measured, but should also be relatively inert. Thus, noble gases such as argon or krypton are often employed. These reference samples may be admitted into the source, concurrently with the compound of interest, and all the ionization efficiency curves can then be measured under identical conditions. On some occasions, especially when rather low ionization or appearance energies are to be measured, organic compounds, such as benzene or 2-chloropropane, are more convenient. These substances have lower ionization energies than noble gases and their use reduces the possibility of errors arising due to any nonlinearity in the electron beam energy scale.

There are several procedures for evaluating the ionization or appearance energy from the ionization efficiency curves. Undoubtedly the best of these is the vanishing ion current approach. In this method, the nominal electron beam energies for the onset of ionization are measured. Since the absolute values of the onset energies of the calibrant compounds are known, the unknown ionization or appearance energy may then be deduced by simple addition or subtraction. This process is illustrated in Fig. 2-3, where the unknown quantity is the ionization energy of the methyl radical. The inset in Fig. 2-3 is the foot of the ionization efficiency curve of krypton, which has an ionization energy of 14.00 eV; this information is used to calibrate the electron energy scale. The threshold for the onset of ionization is remarkably well defined and the ionization energy of the methyl radical may be deduced as 9.84 eV.[2]

There are several disadvantages associated with the vanishing ion current approach. The first is that it is often difficult to measure the ionization efficiency curve sufficiently close to the threshold. This problem is especially pronounced when the foot of the curve exhibits a very slow reduction to zero; this phenomenon

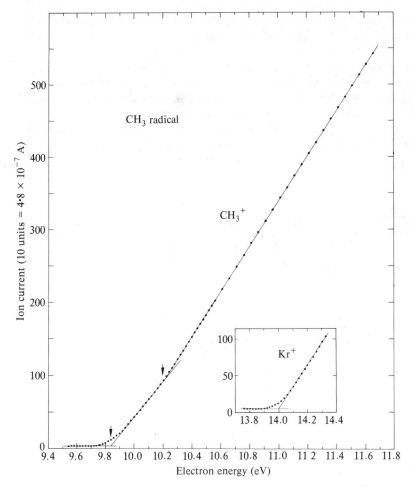

Figure 2-3

is called tailing. A second difficulty arises because of the inhomogeneity of the electron beam energy. The electrons, used to induce ionization, do not have a single well-defined energy, but usually a spread of more than 1 eV. Thus, even at an average electron beam energy below the ionization energy of interest, some electrons may still have enough energy to cause ionization. This problem may be surmounted by using an energy-resolved electron beam to ionize the sample and standard. It is possible to obtain accurate ionization and appearance energy measurements using mass spectrometers equipped with monoenergetic electron beams.[2] Signal-averaging techniques are also employed in order to improve sensitivity and increase the signal-to-noise ratio near the threshold. However, although this method is accurate,[3,4] it requires sophisticated instrumentation. Therefore, it cannot be used with any reliability on conventional mass spectrometers, which do not have monoenergetic electron beams. Furthermore, when using mass spectrometers with standard source and collector fittings, it is difficult, if not impossible, to decide where the threshold for ionization occurs.

The only way to obtain accurate ionization and appearance energies, by electron impact, is to use an energy-resolved electron beam. Nevertheless, despite the difficulties associated with the use of conventional mass spectrometers,[3,4] approximate ionization and appearance energies may be obtained.

A method for analyzing the data, used to produce the ionization efficiency curves, which is suitable for use in deriving such approximate values, is the semilog plot procedure.[5] In practice, the ion current (I), at a given electron beam energy,

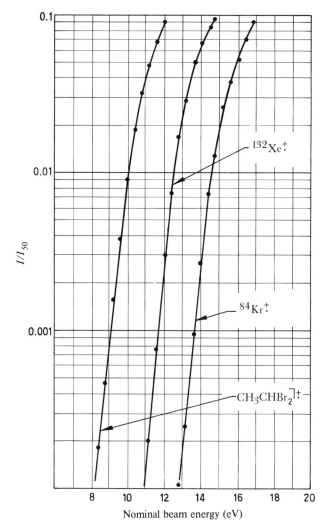

Figure 2-4

is normalized to the ion current (I_{50}) at a nominal electron beam energy of $50\,eV$. The values of I/I_{50}, thus obtained, are plotted against the nominal electron beam energy on semilog graph paper. This process is performed for the compound of interest and the reference substances. It is usually found that the resulting curves are parallel, to a good approximation, in the region where I/I_{50} is 10^{-3} to 10^{-2}. The difference, in electronvolts, between the curves for the substance of interest and the internal calibrant is then taken to be the difference between the ionization (or appearance) energies concerned.

An example of the application of this method is given in Fig. 2-4; the quantity to be measured is the ionization energy of 1,1-dibromoethane and the internal standards are xenon and krypton. The ionization energy of 1,1-dibromoethane is estimated to be $2.6\,eV$ less than that of xenon. The ionization energy of xenon is $12.13\,eV$; consequently, the ionization energy of 1,1-dibromoethane is found to be $9.5\,eV$. A similar calculation, using krypton as reference, yields a value of $9.4\,eV$, which is in satisfactory agreement.

There are several disadvantages associated with the use of semilog plots to determine ionization or appearance energies. No attempt is made to avoid the problems which arise because the ionizing electrons are not monoenergetic. Moreover, the measurements are not made close to the threshold for ionization. Thus, it must be assumed that the ionization efficiency curves are very similar in shape for the compound under investigation and the reference substances. Finally, the normalization procedure is essentially arbitrary; there is no particular reason for performing this normalization with respect to the ion current at an electron beam energy of $50\,eV$. Consequently, systematic errors may be involved, in

Table 2-1 Ionization energies (eV) of some monofunctional aromatic compounds C_6H_5X

X	Photoionization	Electron impact	Photoelectron spectroscopy
NO_2	9.92	10.18	10.26
CN	9.71	10.09	10.02
CHO	9.53	9.70	9.80
$COCH_3$	9.27	9.57	...
H	9.24	9.56	9.40
F	9.20	9.73	9.50
Cl	9.07	9.60	9.31
Br	8.98	9.52	9.25
CH_3	8.82	9.18	8.9
I	8.72	9.27	8.78
OH	8.50	9.16	8.75
SH	8.33	...	...
OCH_3	8.22	8.83	8.54
NH_2	7.70	8.32	8.04
$NHCH_3$	7.35	...	7.73
$N(CH_3)_2$	7.14	7.95	7.51

Table 2-2 Ionization energies (eV) of some mono-functional aliphatic compounds $CH_3CH_2CH_2X$

X	Ionization energy	X	Ionization energy
H	11.07	OH	10.17
ONO_2	11.07	CHO	9.86
Cl	10.82	$CH{=}CH_2$	9.5
CH_3	10.63	$COCH_3$	9.34
OAc	10.54	I	9.26
C_2H_5	10.34	NH_2	8.78
Br	10.18		

addition to random errors, when ionization or appearance energies are evaluated using this procedure. Notwithstanding these comments, the semilog plot method constitutes a relatively simple and rapid process for obtaining approximate values for ionization and appearance energies. It should always be recognized that heats of formation, or transition state energies, derived from these measurements, are only approximate. Even in the most favorable cases, errors of 0.1 eV are almost certainly involved and, in unfavorable situations, larger errors may be present. On the other hand, such errors ought not to mask larger differences in heats of formation, or activation energies for dissociation, which may vary from 0 to 4 eV.

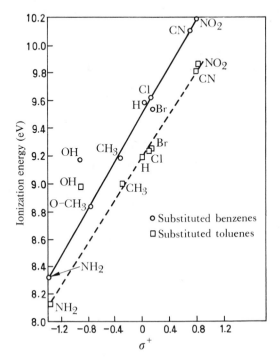

Figure 2-5

The ionization energies for a number of simple monofunctional aromatic and aliphatic compounds are given in Tables 2-1 and 2-2. The values obtained by three methods—photoionization, electron-impact, and photoelectron spectroscopy—are compared in Table 2-1; the values in Table 2-2 are photoionization results.

Simple relationships are found between the ionization energy and the molecular structure. The simplest correlation is found for the monosubstituted aromatic compounds C_6H_5X (Table 2-1). These data show that substituents which have strong electron-withdrawing mesomeric effects (NO_2, CN, CHO, $COCH_3$) increase the ionization energy relative to benzene. In contrast, those substituents which have the largest electron-donating effects [OH, SH, OCH_3, NH_2, $NHCH_3$, $N(CH_3)_2$] cause a reduction in the ionization energy relative to benzene. A good correlation is found between the ionization energy and the σ^+ value for the substituent (Fig. 2-5)[6] in the cases of both monosubstituted benzenes and substituted toluenes.

A relationship also exists between the number of π electrons in a polynuclear aromatic hydrocarbon and its ionization energy.[7] In general, there is a smooth decrease in the ionization energy of a series of polynuclear aromatic hydrocarbons, of the same symmetry, as the number of π electrons is increased. For example, Fig. 2-6 illustrates how the ionization energies of the members of the homologous series of polynuclear aromatic hydrocarbons, beginning with benzene and ending with pentacene, range smoothly from 9.38 to 6.55 eV.

In the series of aliphatic compounds $CH_3CH_2CH_2X$ (Table 2-2) it may be seen that all substituents (with the exception of nitrate) lower the ionization energy with respect to the case where X = H. When the so-called nonbonding electrons are least tightly bound (e.g., in amines), then the effect of the substituent on lowering the ionization energy is greatest. Thus in amines the lowest energy ionization process occurs by removal of a nonbonding electron from nitrogen, and at these energies it is physically significant to regard the charge as being localized on the nitrogen atom (1). At the other extreme, in hydrocarbons there is no single site at which the charge resides with great preference. Here ionization occurs by removal of an electron from a molecular orbital distributed over the entire molecule, so that in an *n*-alkane all the bonds are weakened simultaneously.

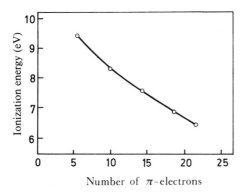

Number of π-electrons   **Figure 2-6**

However, calculations[8] suggest that the positive charge is not distributed completely randomly, but that the charge density in some bonds is greater than in others. Thus, the best notation for the pentane molecular ion is **2**. It will be seen subsequently that notations such as **1** are very useful in understanding the reactions of lowest activation energy of the molecular ion.

$$CH_3CH_2CH_2\overset{+\cdot}{N}H_2 \qquad\qquad CH_3CH_2CH_2CH_2CH_3{}^{\ddagger}$$

$$\textbf{1} \qquad\qquad\qquad\qquad\quad \textbf{2}$$

The spread of values obtained by different methods for the ionization energies of monosubstituted benzenes may be compared. Photoionization results are obtained by ionizing the molecule with a continuously variable and nearly monoenergetic beam of photons.[9] In the photoelectron spectroscopy experiments,[10] ionization is effected by irradiation with a beam of 21.21 eV photons (from a helium resonance lamp). Since the photon producing ionization must give up all its energy (in contrast to an electron), the kinetic energy of the electrons knocked out of the highest occupied molecular orbital will be 21.21 − IE (ionization energy) eV. The kinetic energies of the photoelectrons may be measured with a magnetic-field deflection analyzer, the kinetic energy scale being derived from the ionization energies of the calibrating gases argon, xenon, and krypton. Each of the three methods compared (Table 2-1) are considered to give vertical ionization energies, although in some cases the $0 \to 0$ transition may occur with sufficient probability to be detected. While there is usually a good degree of consistency within the values obtained by any one method, it is apparent (Table 2-1) that the spread of values for any one compound is about 0.3–0.6 eV (with the exception of dimethylaniline), if all three sets of values are considered. Electron-impact values lose some precision because the bombarding electrons are not homogeneous in energy; moreover, the neutral molecules usually acquire a significant amount of thermal energy in the source prior to ionization. Using sophisticated instrumentation, electron-impact determinations may be reproducible within ± 0.1 eV. However, in general, unless comparison with values obtained by other methods can be made, it would seem unwise to rely on the absolute value to better than ± 0.3 eV. Of the values listed in Table 2-1, those

Table 2-3 Some second ionization energies

Molecule	Second IE, eV
Ne	62.5
NH_3	36.8
C_6H_5D (benzene)	26.0
$C_{10}H_8$ (naphthalene)	22.8
$C_{22}H_{14}$ (pentacene)	19.6

obtained by photoelectron spectroscopy are probably the most reliable vertical ionization energies.

When sufficient energy is transferred to the neutral molecule by electron (or photon) impact, a doubly charged molecular ion (M^{++}) may be generated and recorded at an m/z value one-half of that for $M^{\ddagger}$. The minimum energy required to generate M^{++} is known as the second ionization energy and examples are shown in Table 2-3. It is obviously undesirable that two centers of repulsive positive charges should be confined in close proximity and it is not surprising to find that doubly charged molecular ions are only really important in large aromatic compounds.

B. Energy Distributions in Molecular Ions

The excess internal energy, which is present in a molecular ion after ionization by electron impact, is dependent mainly on the energy transferred during the ionization process. The parent molecule also possesses thermal energy before ionization; however, this internal energy component is usually small.

Ionization efficiency curves for the production of positive ions from polyatomic molecules rise rapidly with increasing electron beam energy, for a few electronvolts above the ionization threshold (see Fig. 2-2). Hence, the presence in the electron beam of electrons with energies in excess of the ionization energy causes a rapid rise in the rate of ion formation. Such electrons can promote ionization in several ways; consider a general case where an electron may be ejected from more than one occupied molecular orbital (Fig. 2-7).

The three highest occupied molecular orbitals of the neutral molecule are represented by A. Interaction with an electron having an energy equal to, or only marginally greater than, the ionization energy may only result in the formation of the molecular ion $M^{\ddagger}$ in its ground state B. A more energetic electron may be able to generate the electronically excited molecular ion C. As the energy of the

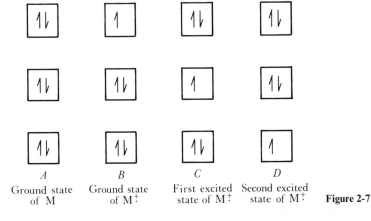

A
Ground state of M

B
Ground state of $M^{\ddagger}$

C
First excited state of $M^{\ddagger}$

D
Second excited state of $M^{\ddagger}$

Figure 2-7

bombarding electrons is increased further, higher excited states of M^+ can also be populated. However, the interaction between the bombarding electron and the neutral molecule may vary. A high-energy electron which interacts relatively weakly with the molecule being ionized may transfer only as much energy as would a low-energy electron which interacted strongly. When a nominal electron beam energy of, say, 20 eV is employed, a population of ions is produced with a distribution of excess internal energies. At one extreme of this distribution, the ions are merely ionized and contain no electronic excitation energy and almost no vibrational energy. At the other extreme, the ions are formed from neutral molecules, which were highly thermally excited, that acquired 20 eV from an electron during ionization. The probability of the occurrence of either of these extremes of behavior is very low. Nevertheless, in terms of the internal energy of the ions produced, the formal limits for the internal energy distribution are 0 to $E_{el} + E_{th} - IE$, where E_{el} is the nominal electron beam energy and E_{th} is the maximum initial thermal energy of the neutral molecules.

It would clearly be of great interest to know the internal energy distribution of the molecular ions (i.e., the percentages of ions having specific energies). This information is not readily available for electron-impact ionization; however, it is available for photoionization.[10]

The kinetic energies of all the electrons, expelled from a molecule upon

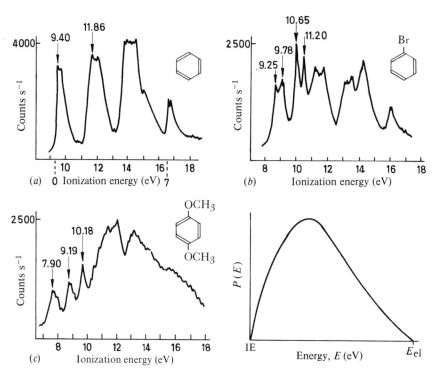

Figure 2-8

irradiation with 21.21 eV photons, may be determined. The differences between 21.21 eV and the measured kinetic energies yield the ionization energies for the removal of electrons from the various molecular orbitals. These data constitute a photoelectron spectrum; several examples of such photoelectron spectra are given in Fig. 2-8. In the case of benzene (Fig. 2-8a), the upper energy scale gives the ionization energies for removal of an electron from the various molecular orbitals; the lower energy scale is referred to the ground state of the benzene molecular ion as zero. Consider, for example, the band at 11.86 eV. This corresponds to formation of benzene molecular ions which initially have 2.46 eV of internal energy (this is mainly electronic energy) above the ground state. The various bands are broadened owing to vibrational fine structure, which is incompletely resolved. The relative abundances of the bands are proportional to the numbers of photoelectrons with a given energy; therefore, these abundances are also proportional to the numbers of molecular ions with a given internal energy. Hence, Fig. 2-8a gives the energy distribution of the benzene molecular ions which are produced by photoionization.

When less symmetrical and more complex molecules are studied, the numbers of molecular orbitals of different energy levels become larger. As a result, the energy distributions are more complicated and less fine structure is clearly resolved.

For example, the photoelectron spectrum of bromobenzene (Fig. 2-8b) exhibits several additional bands compared to benzene. The bands at 10.65 and 11.20 eV are of particular interest; it has been suggested[10] that these bands correspond to removal of an electron from the $4p_x$ and $4p_y$ "nonbonding" orbitals, respectively, of the bromine atom. The p_y orbital, perpendicular to the C—Br bond axis and to the plane of the benzene ring, overlaps more effectively with the π molecular orbitals of benzene than does the p_x orbital which lies in the plane of the benzene ring. Consequently, less energy is needed to remove an electron from the p_x orbital than the p_y orbital.

The photoelectron spectrum of a molecule such as p-dimethoxybenzene (Fig. 2-8c), which is still quite small and symmetrical, is relatively poorly resolved; only a little fine structure can be discerned. Nevertheless, it is evident from Fig. 2-8c that when p-dimethoxybenzene is ionized by photon impact, there is a low probability of producing a molecular ion containing 10 eV of excess energy.

Conclusions concerning the energy distributions of ions formed by photoionization are not necessarily applicable to ions formed by electron impact. The energy levels which may be populated are the same for both photon and electron impact; however, the probabilities associated with the excitation processes are different. Nevertheless, when 21.21 eV photoionization and 20 eV electron-impact mass spectra of complex molecules are compared, gross differences are not usually found. Indeed, semiquantitative calculations of 20 eV mass spectra may be performed, using an assumed energy distribution such as that given in Fig. 2-8d. This energy distribution, which is essentially a smoothed version of that found for photoionization of p-dimethoxybenzene (Fig. 2-8c) or similar molecules, gives fairly good results for the mass spectra of complex molecules.

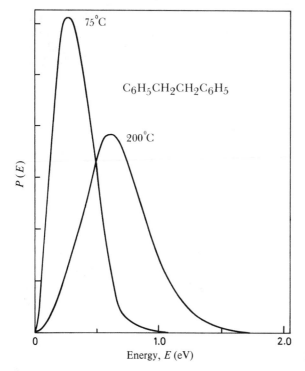

$C_6H_5CH_2CH_2C_6H_5$

Figure 2-9

In common with photoionization results, there is experimental evidence which shows that ionization with 20 eV electrons rarely results in the formation of molecular ions with internal energies of 10 eV or more. In fact, it appears that the most probable energy transfer is between 1 and 8 eV in excess of the ionization energy.[11] Even in 70 eV electron-impact mass spectra, only a small percentage of the molecular ions are formed with energies greater than 10 eV.

The effects of temperature variation on the thermal energy distributions of neutral molecules have also been investigated. For example, the thermal energy distribution for 1, 2-diphenylethane has been calculated for temperatures of 75 and 200°C (Fig. 2-9).[12] Although there is a greater probability of a molecule possessing a larger thermal energy at higher temperatures, even at 200°C few molecules have thermal energies in excess of 1 eV. Consequently, in most instances, the thermal energy of molecules prior to ionization constitutes only a minor part of the excess energy of the molecular ion after ionization.

C. Internal Conversions and Radiative Transitions

When molecular ions are formed with high internal energies, much of the excess energy is present in the form of electronic excitation of the ion which is initially produced. However, as a consequence of the Franck–Condon principle, some ions are also formed as vibrationally excited species within a given electronic state.

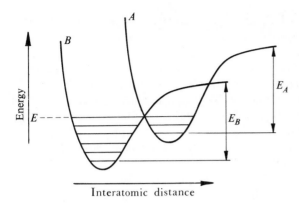

Figure 2-10

Interatomic distance

Dissociation of the molecular ion can occur only when sufficient vibrational quanta are available in the reaction coordinate to provide the activation energy for reaction. Decomposition from any highly vibrationally excited state can take place relatively rapidly (e.g., in times less than 10^{-9} s or even comparable to one bond vibration, that is, 10^{-14} s). Therefore, the mechanisms whereby any electronic excitation energy, available immediately after ionization, may be converted into vibrational energy are important.

Since ions in the source do not interact, intermolecular energy transfer cannot occur. Thus, in the absence of radiative transitions, the total internal energy (E) of each ion would remain constant at the value produced by the initial excitation. Hence, for ions that do not undergo radiative transitions (loss of energy by photon emission), $E = E_{el} + E_{vib} + E_{rot}$, where E_{el}, E_{vib}, and E_{rot} are the electronic, vibrational, and rotation components of the internal energy. In the absence of collisions, E_{vib} and E_{rot} are not interconvertible because the angular momentum of the ion must be conserved; consequently, $E_{el} + E_{vib}$ is constant. The significance of this conclusion may be illustrated by considering two electronic states of a diatomic molecule whose potential energy surfaces cross (Fig. 2-10); similar arguments apply for polyatomic molecules. An equilibrium between these two electronic states is established for ions of internal energy E.

The mechanism whereby the electronic state A is changed into the electronic state B is known as internal conversion. In favorable situations (where the oscillators are anharmonic and efficiently coupled), internal conversions can occur in times corresponding to a vibrational period (10^{-14} s). Since B is the lower-lying electronic state, in this case, internal conversion of A to B causes a conversion of electronic to vibrational energy. As a result, B may have enough vibrational energy to decompose, perhaps very rapidly. However, if decomposition does not occur, B may reconvert to A because there is no mechanism whereby the vibrational energy may be lost. This behavior is in marked contrast to that pertaining for normal thermal and photochemical reactions, where collisions occur and collisional deactivation of vibrationally excited species may take place. Therefore, in the absence of radiative transitions, the amount of reaction occurring from the electronic states A and B depends on the relative activation energies, E_A and E_B, for decomposition from A and B, and the relative populations of the states. Decomposition from the lower-lying electronic state is more probable for two reasons: (1) more internal energy is available in state B than A (alternatively, it may be considered that more internal energy is needed to promote dissociation from state A than B); (2) it is much more probable that a molecule of internal energy E exists in state B, rather than A, because the degeneracy of vibrational states is greater in B. This may be derived by analogy with the case of s similar oscillators containing n quanta, where the number of vibrational states is $(n+s-1)!/n!(s-1)!$. In other words, there are more ways of accommodating the larger number of vibrational quanta in a given number of bonds.

In the general case of a polyatomic molecular ion, many electronic states, together with the associated vibrational states, may be accessible. The number of states per unit of energy is referred to as the density of states; this density of states is usually held to increase rapidly with increasing internal energy. For a high density of electronic states, it is possible to populate low-lying electronic states by internal conversions from higher energy electronic states. The low-lying electronic states, which may perhaps include the ground electronic state of M^{+}, produced in this way are often highly vibrationally excited. Consequently, dissociation may occur relatively rapidly from such highly vibrationally excited states in times of less than 10^{-9} s.

Thus, if there is a low activation energy for decomposition of the molecular ion in the first or second electronically excited state, reaction can take place in times within the range 10^{-14}–10^{-9} s. On the other hand, if the activation energies for dissociation from these states are larger, the lifetimes of ions in the states may be relatively long. Should an ion survive for roughly 10^{-8} s, or longer, energy loss by photon emission is probable, leading to the ground electronic state of the ion. This is because most molecular ions are ion-radicals and usually have a spin (S) of $\frac{1}{2}$; hence, these ions have a spin multiplicity $(2S+1)$ of 2 and exist in doublet states. Since the radiative lifetimes of doublet states are of the order of 10^{-8} s,[13] most molecular ions with lifetimes of this magnitude can undergo radiative transitions to the ground electronic state. However, the intervention of quartet states cannot be ruled out entirely, because double excitation, with spin inversion, is possible during the ionization process.

Some of the possible transitions referred to in this section are summarized in Fig. 2-11. Internal conversions in electronically excited states can, in principle, lead to dense population of the ground state, with molecular ions which are often highly vibrationally excited. Ions in electronically excited doublet states may react from such states within 10^{-8} s, or less; alternatively, they may undergo radiative energy loss and convert to the ground state. Therefore, ions which decompose at relatively slow rates (with a unimolecular rate constant, $k \simeq 10^{6}$ s^{-1}) are likely to react from the ground electronic state. The total internal energy in the molecular ions remains fixed, except that highly excited species may lose energy by radiative transitions.

For complex molecules, no information is as yet available concerning the numbers of electronic states to which excitation may occur. Since the density of

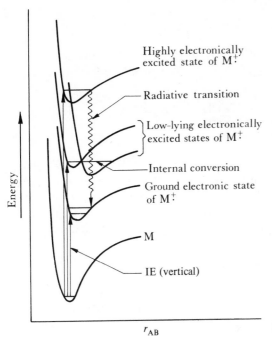

Highly electronically
excited state of M$^{\ddot{+}}$

—Radiative transition

}Low-lying electronically
}excited states of M$^{\ddot{+}}$

—Internal conversion

Ground electronic state
of M$^{\ddot{+}}$

M

— IE (vertical)

Energy

r_{AB}

Figure 2-11

these states generally increases rapidly with increasing internal energy, so the number of crossings of potential energy surfaces also increases rapidly. Furthermore, the vibrations become more anharmonic as the internal energy increases; thus, the vibrational quanta, which initially may be located in certain bonds, can be randomized among other vibrational modes. Consequently, the initial excitation energy, which may be localized in particular parts of the molecular ion, can be rapidly randomized throughout the ion by intramolecular energy transfer. This randomization of a given internal energy E throughout the ion is generally assumed to be fast relative to the rates of subsequent fragmentations.[14]

2-2 CHEMICAL IONIZATION

The arguments developed above concerning the internal energy of the molecular ion after electron-impact ionization apply to low-pressure conditions, where ion-molecule reactions are negligible. However, as shown in Sec. 1-2B, a different situation occurs in a chemical ionization source: ionization of the sample molecules takes place via collision with an excess of reagent ions at high pressure. In this section, the mechanism of chemical ionization is considered, and the current arguments about the internal energy of the ions produced by this alternative ionization process are presented in general terms. An account of the applications of chemical ionization is given in Secs. 7-1 and 7-2.

A. Proton Transfer

Much of the early work on positive-ion chemical ionization employed methane as the reagent gas. At pressures around 1 torr, 90 percent of the total ion current consists of the ions CH_5^+ and $C_2H_5^+$, formed by the following reactions:

$$CH_4 + e \rightarrow CH_4^{\ddagger} + 2e$$
$$CH_4^{\ddagger} \rightarrow CH_3^+ + H\cdot$$
$$CH_4^{\ddagger} + CH_4 \rightarrow CH_5^+ + CH_3 \cdot$$
$$CH_3^+ + CH_4 \rightarrow C_2H_5^+ + H_2$$

The most reactive ion in the methane plasma is CH_5^+ and its relative abundance is a reliable yardstick for attainment of good chemical ionization conditions. For example, an m/z 17 to 16 ratio ($[CH_5^+]$ to $[CH_4^{\ddagger}]$) of 10 to 1 should give excellent chemical ionization. CH_5^+ reacts exothermically with almost all organic molecules, M, behaving as a strong Brönsted acid to yield a protonated molecular ion MH^+:

$$M + CH_5^+ \rightarrow MH^+ + CH_4$$

Over the past ten years many different reagent gases have been investigated and results show that chemical ionization frequently occurs by proton transfer to the sample molecule from an acidic reagent ion:

$$M + BH^+ \rightarrow MH^+ + B$$

The exothermicity of the proton-transfer reaction (directly related to the proton affinity) determines the internal energy of the protonated molecular ion, MH^+, and hence the extent of fragmentation. Therefore, the proton affinity of the conjugate base is an important parameter in determining chemical ionization spectra; a series of proton affinities, each appropriate to a particular reagent gas, is shown in Table 2-4.

The data of Table 2-4 may be used for a simple qualitative prediction of the average internal energy content of MH^+, formed from a variety of reagent ions. For instance, a given protonated molecular ion produced via isobutane ionization is expected to possess less internal energy than that formed in the methane

Table 2-4 Gas-phase proton affinities of some bases B

Reagent gas	Reagent ion	Conjugate base B	Proton affinities (B), $kJ\,mol^{-1}$
H_2	H_3^+	H_2	420
CH_4	CH_5^+	CH_4	530
H_2O	H_3O^+	H_2O	700
$(CH_3)_3CH$	$(CH_3)_3C^+$	$(CH_3)_2C{=}CH_2$	815
NH_3	NH_4^+	NH_3	865
CH_3NH_2	$CH_3NH_3^+$	CH_3NH_2	915

chemical ionization plasma. This is borne out in practice by the observation that methane chemical ionization spectra show more abundant fragment ions than isobutane spectra, for the same sample molecules.

Mixtures of reagent gases can be employed in order to generate additional reagent ions, thus increasing the scope of chemical ionization. For example, the reagent ions N_2H^+, HCO_2^+, and HCO^+ can be formed in high abundance using mixtures of hydrogen with nitrogen, carbon dioxide, and carbon monoxide, respectively.

It appears that more than about $500\,kJ\,mol^{-1}$ of internal excitation in MH^+ is uncommon, even when hydrogen is used as the reagent gas. The consequences of this in terms of the appearance of CI spectra are discussed in more detail in Chap. 7; in the present context, the relative abundance of ions in the molecular ion region of CI spectra is greater than that found for EI spectra. This is caused in part by the different internal energies of the respective MH^+ and $M^{\ddot{+}}$ ions; in addition, the even-electron MH^+ ions possess an inherent stability compared with the radical $M^{\ddot{+}}$ ions.

B. Reagent Ion Capture

The simple (and common) cause of chemical ionization by sample protonation is dealt with above. However, ionization also may occur via bimolecular association reactions which are generally classified as solvation processes, e.g.,

$$M + C_2H_5^+ \rightarrow M \cdot C_2H_5^+ \qquad \text{(minor reaction in methane CI)}$$

$$M + NH_4^+ \rightarrow M \cdot NH_4^+ \qquad \text{(common reaction in ammonia CI)}$$

$$M + NO^+ \rightarrow M \cdot NO^+ \qquad \text{(common reaction in nitric oxide CI)}$$

As indicated above, the NH_4^+ ion, produced by electron bombardment of ammonia at 1 torr, not only functions as a weak Brönsted acid but also as an electrophile toward organic molecules in the gas phase. Thus, carbonyl compounds such as monofunctional ketones, aldehydes, esters, and acids add NH_4^+, but are not sufficiently basic to form MH^+ by proton transfer. However, if the proton affinity of the sample is greater than that of ammonia, MH^+ ions are formed (e.g., amides, amines, α,β-unsaturated ketones). The solvation properties of NH_4^+ are particularly useful in analysis of certain polyfunctional compounds; this is discussed in Chap. 7.

C. Charge Transfer

The general definition of chemical ionization given in Sec. 1-2 also encompasses ionization by charge transfer. In this type of reaction, the sample is ionized merely by transfer of an electron between the sample molecule and the ionized reagent gas:

$$M + X^{\dagger} \rightarrow M^{\dagger} + X \qquad (2-2)$$

The reagent gas is frequently a monatomic species (i.e., one of the noble gases) and seldom contains more than three atoms. In principle, the reagent ion $X^{\dagger}$ is present in its ground state and a single, well-defined amount of energy (known as the recombination energy, RE) is transferred to the sample molecule M in the ionization process. Thus, for example, the recombination energy of argon is 15.8 eV and the $Ar^{\dagger}$ ion would be predicted to form the molecular ion $M^{\dagger}$ having a single internal energy of $(15.8 - IE)$ eV, where IE is the ionization energy of M. The molecular ion would then decompose on the mass spectrometer time scale to an extent governed by this energy; thus, the extent of breakdown of $M^{\dagger}$ may be controlled by the use of different reagent ions. Table 2-5 shows the RE values appropriate to a set of reagent gases. For gases included in Table 2-5, ionization by $He^{\dagger}$ should produce the highest energy molecular ion and NO_2^{+} the lowest.

However, the ideal situation of a monoenergetic molecular ion is not realized in practice. This is because (1) virtually all ions have several different recombination energy values and less internal energy than the maximum predicted may be transferred, and (2) some of the exothermicity of the charge-transfer process may appear as translation. In fact, ionization of polyatomics by charge transfer usually forms molecular ions with a range of internal energies extending below the maximum expected from the recombination energy value.

Charge-transfer mass spectra have no important practical advantages over electron-impact mass spectra as a structural tool; they are used largely for the practical verification of unimolecular rate theories. However, appropriate gas mixtures have been employed in high-pressure ion sources to produce informative mass spectra by initiating two separate reaction channels, one of which involves charge transfer. For instance, using an argon water-vapor mixture, $M^{\dagger}$ ions are produced [by charge transfer to molecules of ionization energy (IE)] which possess up to $(15.8 - IE)$ eV internal energy. These ions undergo extensive decomposition, whereas the MH^{+} ions produced from the H_3O^{+} reagent ions are stable.

Table 2-5 Recombination energies of some charge-exchange reagents

Ion	Recombination energy, eV	Ion	Recombination energy, eV
NO_2^{+}	11.0	CO_2^{+}	14.4
$Xe^{\dagger}$	12.2	N_2^{+}	15.6
$O_2^{\dagger}$	12.5	$Ar^{\dagger}$	15.8
$Kr^{\dagger}$	14.0	$Ne^{\dagger}$	21.6
$CO^{\dagger}$	14.1	$He^{\dagger}$	24.6

D. Negative Ions

It is only during the last few years that negative-ion mass spectrometry has been used to solve structural and analytical problems. Prior to this time, it was recognized that a problem associated with the technique was that many organic compounds do not form stable negative ions. In addition, since high-energy electrons (10–70 eV) had traditionally been used to generate negative ions, the ion currents were often small. The cause-and-effect relationship between high-energy electrons and low ion currents was not widely appreciated.

The interaction of an electron with a sample molecule AB can produce negative ions by three different mechanisms:[15]

$$AB + e \rightarrow AB^{-} \qquad \text{(resonance capture)} \qquad (2\text{-}3)$$

$$AB + e \rightarrow A\cdot + B^{-} \qquad \text{(dissociative resonance capture)} \qquad (2\text{-}4)$$

$$AB + e \rightarrow A^{+} + B^{-} + e \qquad \text{(ion-pair production)} \qquad (2\text{-}5)$$

If normal electron-impact conditions (40–70 eV) are used to produce negative ions, then negative-ion formation occurs predominantly by the ion-pair mechanism. In general, the electron affinity of AB will be far too low to allow capture of an electron with a large kinetic energy to produce AB^{-} by resonance capture. In contrast, ion-pair production [Eq. (2-5)] does provide an outlet for some of this energy through bond cleavage. Nevertheless, under EI conditions, sensitivity is a problem; moreover, fragment rather than molecular ions are usually observed.

The key to efficient formation of negative ions lies in work employing an argon discharge.[16] This method produces a large population of low-energy electrons ("near-thermal" electrons).

A similar population of electrons is also produced under chemical ionization conditions.[17] The highly energetic electrons which are emitted from a filament lose energy in two ways: by promoting positive ion formation (Sec. 2-1) or by colliding with neutral gas molecules. The near-thermal electrons so produced can then undergo resonance capture [Eq. (2-3)], assuming that AB has a positive electron affinity. Under the most favorable conditions, when AB possesses both a positive electron affinity and a large cross section for electron capture, such negative-ion spectra can exhibit a hundred- to thousandfold increase in sensitivity above those found with other ionization techniques.[17] This is thought to arise from the small mass of the electron compared with organic ions. For conditions near thermal equilibrium, the electrons in a CI plasma have much larger velocities than the heavier ions. Consequently, the rate constant for formation of a negative ion by resonance capture can be ca. 400 times the diffusion controlled limit for reaction of an ion with a molecule. On this basis, high-pressure electron-capture ionization [occurring via Eq. (2-3)] ought to be 100 to 1000 times more sensitive than positive-ion CI.

In contrast to electron-capture ionization described above, negative-ion chemical ionization (NICI) may also be used. Reagent ions are formed in the source by electron bombardment of suitable reagent gases; for example, Cl^-, O^-, and CH_3O^-, respectively, can be produced from carbon tetrachloride, nitrous oxide, and methyl nitrite reagent gases. These reagent ions then react with the sample molecules to form negative ions. For instance, CH_3O^- can act as a Brönsted base, producing $[M-H]^-$ ions by abstracting a proton from the sample molecule:

$$AH + CH_3O^- \rightarrow A^- + CH_3OH$$

Alternatively, the reagent ion may add to the sample molecule; thus, Cl^- can add to a carbonyl compound by nucleophilic attack to produce a $[M+Cl]^-$ ion:

Applications of NICI are presented in Sec. 7-2.

2-3 FIELD IONIZATION AND FIELD DESORPTION

The discussion in Secs. 1-2C and 1-2D introduced the concept of ionization by the application of high electric fields to organic samples, either present in the gas phase (field ionization) or adsorbed on an activated anode emitter (field desorption).[18]

Unlike ionization by electron impact, the field ionization and field desorption processes produce organic cations with little excess internal energy. This factor, added to the short times available for fragmentation (usually less than 10^{-10} s; see Sec. 3-4 for relevant discussions), results in spectra which exhibit peaks mainly in the molecular ion region.

As yet there is no satisfactory quantitative theory that describes the field desorption of large organic molecules from carbonaceous microneedles and efficient ion production remains partly an "art." However, the following parameters are considered important in promoting efficient ionization:[18,19]

1. *Emitter heating current.* An optimum emitter current (usually 10–40 mA for activated 10 µm tungsten wires) exists for ionization of each organic molecule. Reproducibility is enhanced by emission-controlled emitter heating.
2. *Structure of emitter.* Optimum results are obtainable with microneedles 20–40 µm in length. Long needles increase the surface area for adsorption of the sample but decrease the field strength in the emission centers; the quoted length is the best compromise.
3. *Choice of solvent.* Different solvents yield dramatically different field desorption spectra on occasions. For example, methanol (usually containing trace amounts of Na^+) frequently facilitates the formation of abundant $M+Na^+$ ions, when

used as the solvent for loading the emitter. Such ions may be used as an aid to mass counting.

4. *Addition of alkali metal salts.* It is sometimes profitable to add (for instance) Na^+ or Li^+ salts to the emitter to promote field desorption. This has been found particularly advantageous in enhancing sensitivity in the spectra of sugars.

The above brief discussion describes some of the most important operations required to promote efficient production of cations in the field ionization or field desorption processes. Some specific examples of the applications of field desorption are given in Sec. 7-3.

REFERENCES

1. J. H. Beynon, R. A. Saunders, and A. E. Williams: "The Mass Spectra of Organic Molecules," Elsevier, Amsterdam, 1968.
2. F. P. Lossing and G. P. Semeluk: *Can. J. Chem.*, vol. 48, p. 955, 1970.
3. J. H. Beynon, R. G. Cooks, K. R. Jennings, and A. J. Ferrer-Correia: *Int. J. Mass Spec. Ion Phys.*, vol. 18, p. 87, 1975.
4. H. M. Rosenstock: *Int. J. Mass Spec. Ion Phys.*, vol. 20, p. 139, 1976.
5. F. P. Lossing, A. W. Tickner, and W. A. Bryce: *J. Chem. Phys.*, vol. 19, p. 1254, 1951.
6. G. F. Crable and G. L. Kearns: *J. Phys. Chem.*, vol. 66, p. 436, 1962.
7. E. J. Gallegos: *J. Phys. Chem.*, vol. 72, p. 3452, 1968.
8. J. Lennard-Jones and G. G. Hall: *Proc. Roy. Soc.*, vol. 213A, p. 102, 1952.
9. K. Watanabe: *J. Chem. Phys.*, vol. 26, p. 542, 1957.
10. A. D. Baker, D. P. May, and D. W. Turner: *J. Chem. Soc. (B)*, p. 22, 1968.
11. See, for example, W. A. Chupka and M. Kaminsky: *J. Chem. Phys.*, vol. 35, p. 1991, 1961.
12. F. W. McLafferty, T. Wachs, C. Liftshitz, G. Innorta, and P. Irving: *J. Am. Chem. Soc.*, vol. 92, p. 6867, 1970.
13. R. C. Dougherty: *J. Am. Chem. Soc.*, vol. 90, p. 5780, 1968.
14. H. M. Rosenstock, M. B. Wallenstein, A. L. Wahrhaftig, and H. Eyring: *Proc. Natl. Acad. Sci. U.S.A.*, vol. 38, p. 667, 1952.
15. D. F. Hunt, G. C. Stafford, Jr., F. W. Crow, and J. W. Russell: *Anal. Chem.*, vol. 48, p. 2098, 1976.
16. M. von Ardenne, K. Steinfelder, and R. Tummler: "Electronenanlagerungs-Massenspektrographie Organischer Substanzen," Springer-Verlag, New York, 1971.
17. D. F. Hunt and F. W. Crow: *Anal. Chem.*, vol. 50, p. 1781, 1978.
18. H. D. Beckey: "Principles of Field Ionization and Field Desorption Mass Spectrometry," Pergamon Press, Oxford, 1977.
19. H.-R. Schulten and N. M. M. Nibbering: *Biomed. Mass Spec.*, vol. 4, p. 55, 1977.

THREE

KINETICS OF UNIMOLECULAR REACTIONS

Three general classes of ions may be detected using mass spectrometers equipped with magnetic analyzers; these are molecular ions, daughter ions, and metastable ions. Molecular ions do not decompose and consequently are transmitted intact through the mass spectrometer. In contrast, normal daughter ions arise by fragmentation of the molecular ion, or by decomposition of other daughter ions, in the source; these ions are then transmitted through the instrument. However, it is possible that an ion may dissociate after leaving the source but before reaching the collector. Ions which behave in this way are called metastable ions and reactions arising from the dissociation of such ions are referred to as metastable transitions. Molecular ions and daughter ions are commonly encountered in all types of mass spectrometer; metastable ions are most conveniently observed using magnetic-sector instruments.

The relative abundances of the molecular, daughter, and metastable ions is an important feature of the mass spectrum. There are two important parameters in determining the extent to which fragmentation reactions occur. The first of these is the amount of internal energy available in the molecular ion, or fragment ion, to promote dissociation; this effect has been discussed in the previous chapter. The second parameter is the time taken for ions to leave the source, traverse the instrument, and arrive at the collector. This time factor is important because the mass spectrum originates from the kinetically controlled decomposition of the molecular ion. Therefore, it is necessary to consider the rate constants appropriate to dissociations which take place in the various regions of the mass spectrometer. These rate constants depend on the time spent by the ions in the source and on the time needed to reach other parts of the instrument. In this chapter, the importance of these rates in influencing the appearance of the mass spectrum is discussed.

3-1 ION SOURCE AND ANALYZER RESIDENCE TIMES

In representative commercial mass spectrometers currently in use for organic analysis, ions typically spend somewhere in the region of 1–5 μs in the source. Unfortunately, it is difficult to define this time precisely; however, it is known to be reduced by increasing the ion repeller voltage or the accelerating voltage. Ions are repelled more effectively from the source with larger repeller potentials; in addition, since the accelerating voltage penetrates into the source, the ions tend to leave the source more rapidly as the accelerating voltage is increased.

Fortunately, once the ion leaves the source and is accelerated, the problem becomes simpler. The overall length of the ion trajectory is relatively well defined and, provided that the velocity of the ion is known, reasonable estimates can be made for the time taken to reach the collector. Equally, the time of flight which must elapse before an ion passes through the electric sector and enters the second field-free region may also be calculated.

For an ion of mass m and velocity v which has been accelerated through a voltage difference V, from Eq. 1-3:

$$v = \left(\frac{2zeV}{m}\right)^{1/2} \qquad (3\text{-}1)$$

Although the velocity of the ion changes because the ion is deflected by the electric and magnetic analyzers, the speed is not greatly affected. Hence, the time (t) taken to traverse a given distance (x) is given approximately by

$$t = x \left(\frac{m}{2zeV}\right)^{1/2} \qquad (3\text{-}2)$$

Thus, the time taken for an ion to arrive at a given part of the mass spectrometer, after acceleration, is proportional to $m^{1/2}$ and $V^{-1/2}$.

For example, in the AEI MS 9 double-focusing mass spectrometer,[1] the times required for ions to reach various parts of the instrument have been calculated.[2] At an accelerating voltage of 8 kV, an ion of m/z 100 takes approximately 15 μs to reach the collector; about half of this time is spent reaching the second field-free region. At 2 kV, these times are roughly doubled, 30 μs elapsing before the ion arrives at the collector. These times are also dependent on the mass of the ion; nevertheless, the times needed to reach the end of the first field-free region from the source are generally slightly longer than the source residence times (i.e., say 2–5 μs at 8 kV).

The problem of relating these approximate times to rate constants is slightly complicated because ions of a particular internal energy decompose with a certain rate constant, which is associated with a spread of ion lifetimes. Suppose that the size of the vibrational quantum is a certain single value in a polyatomic molecule, which contains a total internal energy corresponding to 10 such quanta. Let the requirement for decomposition be that 5 quanta must be gathered into the reaction coordinate. Decomposition occurs whenever any 5 of the 10 quanta fluctuate into the reaction coordinate; this is a statistical problem. Clearly, there

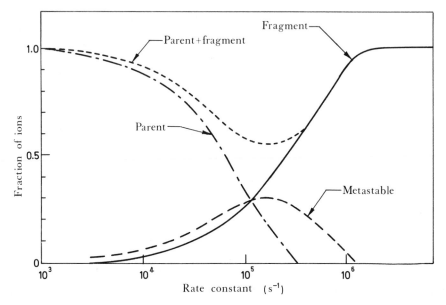

Figure 3-1

will be a most probable time which must elapse before decomposition occurs; however, in some molecules (of the same internal energy) reaction takes place after a somewhat shorter or longer time. Thus, there is a finite possibility that an ion in a population which decomposes with a unimolecular rate constant (k) of $10^5 \, \text{s}^{-1}$ (i.e., a mean lifetime of $10 \, \mu\text{s}$) can survive for as little as $1 \, \mu\text{s}$ or as long as $100 \, \mu\text{s}$. These two extremes of behavior would give rise to a normal daughter ion formed in the source and to an intact parent ion; decomposition after $10 \, \mu\text{s}$ would probably give rise to a metastable ion in the second field-free region. The overall situation is summarized in Fig. 3-1.

Figure 3-1 shows quite clearly that ions decomposing with a rate constant of $10^5 \, \text{s}^{-1}$ contribute significantly to the parent and fragment ions as well as to the metastable ions. The rate constants contributing to the metastable transitions lie, to a good approximation, in the range 10^4–$10^6 \, \text{s}^{-1}$. An ion decomposing with a rate constant greater than $10^6 \, \text{s}^{-1}$ gives rise virtually exclusively to normal daughter ions, while those ions with rate constants of less than $10^4 \, \text{s}^{-1}$ are almost all recorded as parent ions.

Several important conclusions emerge from the discussion above and from a consideration of Fig. 3-1. For instance, the detection of a parent ion is usual only if the rate constant associated with decomposition of this ion is less than about $10^4 \, \text{s}^{-1}$. This is a very slow rate compared to the time scale of molecular bond vibrations ($v \sim 10^{14} \, \text{s}^{-1}$). Any ion which does not have enough energy to decompose must be recorded as a parent ion. However, even some of the ions that have enough energy to decompose do not do so before arriving at the collector. Thus, nondecomposing ions do not necessarily have insufficient energy to dissociate. On the other hand, a fragment ion produced by decomposition of the molecular ion can only be detected if the associated rate constant is roughly $10^4 \, s^{-1}$ or more. Rate constants of ca. $10^5 \, s^{-1}$ correspond to dissociation of metastable molecular ions in the second field-free region. For faster rates, for example, $10^6 \, s^{-1}$ or greater, decomposition can occur with high probability in the source; this gives rise to a normal fragment ion. In general, the slowest reaction rates correspond to the lowest average internal energies sufficient to promote reaction. Metastable ions are those which have the slowest rate constants appropriate for decomposition in the mass spectrometer. Therefore, metastable ions are those which possess internal energies only just in excess of the activation energy for dissociation. Ions with greater excess energies decompose at faster rates in the source.

3-2 METASTABLE IONS

Metastable ions dissociate during their passage from the source to the collector. In double-focusing mass spectrometers, there are three field-free regions (sometimes called drift regions) which the ions traverse after acceleration but before arriving at the collector. The names of these regions arise because ions in them are not acted on by magnetic or electric fields; as a result, the ions drift through these regions. These field-free regions are situated before the first analyzer, between the two analyzers, and after the second analyzer; they are referred to as the first, second, and third field-free regions, respectively. Figure 3-2 shows the positions of the three field-free regions (denoted by $^1m^*$, $^2m^*$, and $^3m^*$ regions) of a double-focusing mass spectrometer, in which ions are transmitted by the electric-sector analyzer (ESA) before entering the magnetic analyzer (MA). Metastable ions which dissociate in the first, second, and third field-free regions give rise to first, second, and third field-free region metastable peaks, respectively; metastable peaks are usually denoted by asterisks, i.e., $^1m^*$, etc. Metastable peaks arising from decompositions in each of the three field-free regions may be detected in principle. In practice, for instrumental reasons, metastable transitions are usually observed in either the first or second field-free regions.

A. Metastable Transitions in the Second Field-Free Region

An ion of charge ze possesses a kinetic energy zeV after being accelerated through a potential difference of V volts. Consequently, all ions which enter the first field-free region, after acceleration, have the same translational energy, zeV, which is independent both of the mass of the ion and of the internal energy of the ion. The initial Boltzmann distribution of translational energies and field inhomogeneities in the source cause a slight velocity spread, which is significant when trying to achieve very high resolving powers (Sec. 1-4). However, in the present context, this velocity spread is insignificant for kinetic energies of several thousand electronvolts. For an instrument having the basic geometry depicted in Fig. 3-2,

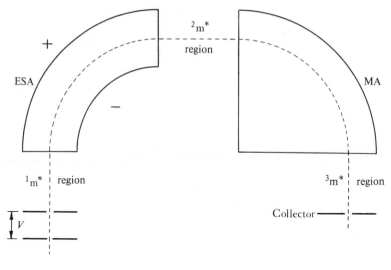

Figure 3-2

the voltage, E, applied between the radial plates of the electrostatic analyzer is tuned so as to transmit only ions of kinetic energy zeV. This is a general feature of normal mode operation with such instruments: V and E are always adjusted so that V/E remains constant, irrespective of whether 2, 4, 6, or 8 kV spectra are being determined. Thus, the accelerating voltage is said to be coupled to the electrostatic analyzer voltage.

Hence, ions of energy zeV arrive in the second field-free region. However, if the reaction $m_1^+ \rightarrow m_2^+$ occurs in the second field-free region, then because the precursor ion m_1^+ possessed the energy zeV the daughter ion m_2^+ will possess the kinetic energy $(m_2/m_1)\,zeV$. This follows since the kinetic energy is partitioned in the ratio of the masses of the two product species m_2^+ and $m_1 - m_2$, the neutral species carrying off $[(m_1 - m_2)/m_1]\,zeV$. Therefore, m_2^+ ions formed in the second field-free region possess less kinetic energy than "normal daughter" m_2^+ ions formed in the source; consequently, they are deflected more readily when traversing the magnetic analyzer. The ions give rise to a peak in the normal mass spectrum at a position $^2m^*$ on the mass scale which is given by $m_2 \times m_2/m_1$,[3] i.e.,

$$^2m^* = m_2^2/m_1 \tag{3-3}$$

Metastable peaks can also be observed in the normal mass spectrum using single-focusing instruments which are not equipped with electrostatic analyzers. These peaks are formed when m_1^+ decomposes into m_2^+ in the first field-free region—in other words, after acceleration but before entering the magnetic analyzer. In general, remarks concerning metastable transitions occurring in the field-free region immediately preceding the magnetic analyzer are equally applicable for single- and double-focusing instruments. This generalization holds because the electrostatic analyzer in double-focusing instruments which have the basic

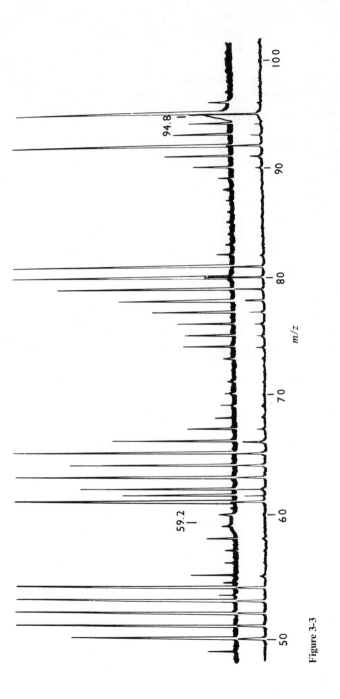

Figure 3-3

geometry depicted in Fig. 3-2 functions as an energy focusing device. The electrostatic analyzer ensures that all ions which enter the second field-free region preceding the magnetic analyzer have the kinetic energy zeV. A special case arises in "reversed geometry" instruments in which ions are transmitted by the magnetic analyzer before entering the electrostatic analyzer. Such machines may be used effectively as single-focusing instruments by observing dissociation of ions in the first field-free region, which is evidenced by metastable peaks in the normal mass spectrum.

Metastable peaks are usually readily identifiable in the normal mass spectrum because they (1) need not occur near integral m/z values, (2) are nearly always much broader than normal peaks, and (3) are of relatively low abundance. On a scale that is roughly linear in m/z values, it is useful to remember that, for the reaction $m_1{}^+ \rightarrow m_2{}^+$, m* lies a distance below m_2 on the mass scale which is similar to the distance that m_2 lies below m_1. This relationship becomes exact when a logarithmic scale of m/z values is employed. The abundance of m* is frequently only of the order of 10^{-2} (or less) of the abundance of the parent and daughter ions in 70 eV spectra; this is because it is usual for many more ions to react in the source than in the second field-free region.

Two metastable peaks in part of the spectrum ($m/z = 50$–100) of p-amino-anisole are evident in Fig. 3-3. The assignments of the metastable peaks are as shown in Table 3-1. The peak at m/z 94.8 is roughly gaussian in shape and partially overlaps with the *normal* peak at m/z 95, while that at m/z 59.2 is much broader and flat-topped.

For the vast majority of cases where metastable transitions are observed, there is no reasonable doubt that the reaction $m_1{}^+ \rightarrow m_2{}^+$ occurs in one step, but the position of the metastable peak is independent of the number of steps. A documented example of a two-step metastable transition ($C_7H_7{}^+ \rightarrow C_3H_3{}^+$) is

Table 3-1 Assignments of metastable peaks in the partial spectrum of p-aminoanisole

^{2}m*	Transition	Inferred reaction
m/z 94.8	m/z 123 → 108	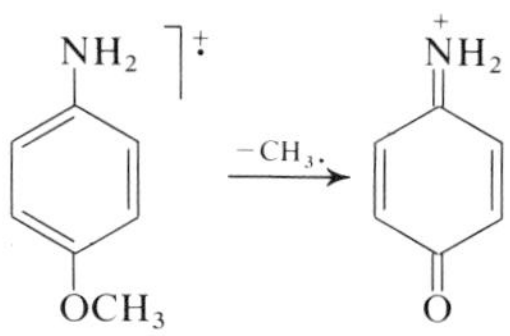
m/z 59.2	m/z 108 → 80	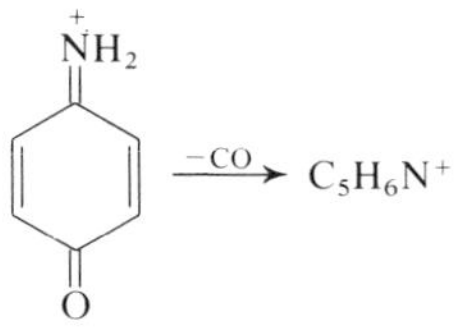

found in the spectrum of toluene, where $C_5H_5^+$ ions formed from $C_7H_7^+$ ions in the first drift region were specifically transmitted by the electrostatic analyzer (see part B of this section) and found to decompose to $C_3H_3^+$ ions in the second drift region.[4] Many instances of consecutive collision-induced metastable transitions are also known (see Sec. 8-4).

Although the vast majority of metastables result from decompositions of singly charged species, examples of metastable decompositions of doubly charged ions are known – to form either two singly charged ions or a doubly charged ion plus a neutral. An example of the former is found in the mass spectrum of benzene,[5] where the metastable transition $C_6H_6^{++} \rightarrow CH_3^+ + C_5H_3^+$ occurs, as evidenced by broad metastable peaks centered at $m/z = 5.77$ and 101.8. Since doubly charged ions possess twice the kinetic energy of singly charged ions after acceleration, it may readily be shown that for the reaction $m_1^{++} \rightarrow m_2^+ + m_3^+$, *two* metastable peaks will occur at m/z values of $2m_2^2/m_1$ and $2m_3^2/m_1$. In the above example, one of the metastables is found at a higher mass than the molecular ion.

B. Metastable Transitions in the First Field-Free Region

Consider a double-focusing mass spectrometer with the basic geometry depicted in Fig. 3-4; suppose an ion m_1^+ decomposes into m_2^+ (closed circles) and an associated neutral species (open circles), after acceleration but before entering the electrostatic analyzer. The electrostatic analyzer will not transmit m_2^+ because it possesses a kinetic energy $(m_2/m_1)zeV$. However, m_2^+ ions formed exclusively in the first field-free region can be transmitted at the pre-set electrostatic analyzer voltage of E if their kinetic energy is increased in the ratio m_1/m_2. To do this, the accelerating voltage must be increased from its normal value V_0 to V_1, such that

$$\frac{V_1}{V_0} = \frac{m_1}{m_2} \tag{3-4}$$

whilst keeping the electrostatic analyzer voltage constant. Any daughter ions,

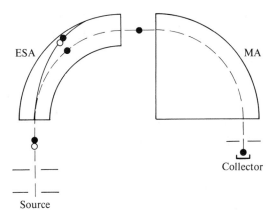

ESA

MA

Collector

Source

Figure 3-4

m_2^+, formed by decomposition of m_1^+ in the first field-free region, enter the electrostatic analyzer with a kinetic energy $(V_1/V_0)zeV \times (m_2/m_1) = zeV$; consequently, they are transmitted through the mass spectrometer (broken line in Fig. 3-4). However, any ions produced in the source (for example, m_1^+) acquire a greater kinetic energy of $(V_1/V_0)zeV = (m_1/m_2)zeV$ because they have experienced an accelerating voltage of V_1/V_0. Therefore, these ions possess too much kinetic energy to pass through the electrostatic analyzer. As a result, they are deflected insufficiently by the electric field and impinge upon the outer wall of the electrostatic analyzer (solid line in Fig. 3-4). Thus, the increase of the accelerating voltage, from V_0 to V_1 at constant E, prevents transmission of all ions formed in the source.[6]

If V_0 were set at 8 kV, then V_1 would be greater than 8 kV; unfortunately, accelerating voltages much in excess of 8 kV can cause breakdown in the source by sparking. Therefore, it is usual to operate with V_0 set initially at 2 or 4 kV.[7] In practice, the mass spectrometer is set up for normal, say 2 kV, operation and the magnet current is adjusted so that the m_2^+ ion of interest arrives at the collector. The accelerating voltage V is then scanned slowly toward higher values; m_2^+ is only transmitted to the collector again when $m_1^+ \rightarrow m_2^+$ occurs in the first field-free region and the accelerating voltage V is such that Eq. (3-4) is satisfied. Since m_2^+ is known and the ratio V_1/V_0 is measured experimentally, it is possible to determine all values of m_1^+ ions which are precursors of m_2^+ ions in the first field-free region. This method of observation of metastable transitions is known as the *metastable refocusing* technique, or the *high-voltage* (HV) scan method.

For example, all precursors of m/z 43 ($C_3H_7^+$ and/or $CH_3C^+{=}O$) in the mass spectrum of methyl n-pentyl ketone (1) can be determined by this method and are summarized in Fig. 3-5a, where the relative abundances of the various transitions are also indicated. If a similar experiment is carried out on the d_5-ketone (2), then the precursors of $CD_3C^+{=}O$ (m/z 46) alone can be determined (Fig. 3-5b).

From Fig. 3-5b it is evident that 2^+ decomposes directly to CD_3CO^+ (m/z 46) or via $[M-CH_3]^+$, $[M-C_2H_5]^+$, and $[M-C_3H_7]^+$ ions (see 2). In addition, $[M-C_3H_6]^+$ (m/z 77) and $[M-C_4H_8]^+$ (m/z 63) ions act as precursors of CD_3CO^+. The main difference between Fig. 3-5a and Fig. 3-5b (apart from the expected mass shifts) is the large abundance of the $71 \rightarrow 43$ transition in (a) compared to the analogous $76 \rightarrow 46$ transition in (b). This arises because the major portion of the $71 \rightarrow 43$ reaction is associated with the step $C_5H_{11}^+ \rightarrow C_3H_7^+ + C_2H_4$ and for the $C_3H_7^+$ ion, no shift of m/z 43 to m/z 46 is possible upon deuteriation.

There are several advantages associated with observing metastable transitions in the first field-free region: (1) each transition is unequivocally identified in terms of the mass of parent and daughter ions; (2) all the precursor ions of m_2^+ may rapidly be found in one scan of the accelerator voltage; (3) normal ions do not interfere; and (4) in the absence of normal ions, the electron multiplier (collection sensitivity) can be increased and so weak transitions may be detected (down to abundances corresponding to 1 ppm of the base peak abundance).

In addition to achieving refocusing by increasing the accelerating voltage at a

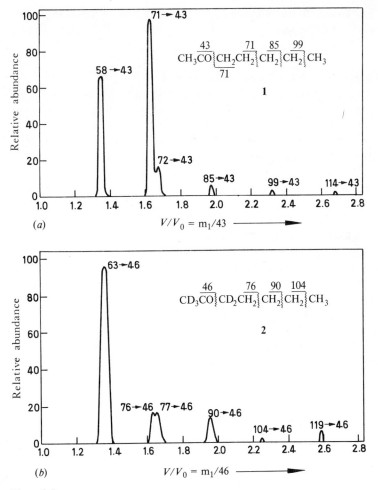

Figure 3-5

given electrostatic analyzer voltage, it is possible to reduce the electrostatic analyzer voltage at a constant accelerating voltage.[8] This alternative refocusing method involves tuning the magnetic-sector voltage to transmit the metastable ions arising from the transition of interest, which are formed in the first field-free region. Thus, the magnetic analyzer is adjusted to transmit m_2^+ ions with a kinetic energy $(m_2/m_1)zeV$. The electrostatic analyzer voltage (E) is now decoupled from the accelerating voltage and reduced from the initial value (E_0) until

$$\frac{E}{E_0} = \frac{m_2}{m_1} \tag{3-5}$$

As a result only ions of kinetic energy $(m_2/m_1)zeV$ are able to pass through the electrostatic analyzer; these ions arise by decomposition of m_1^+ to m_2^+ in the first field-free region.

This technique has been used to demonstrate that doubly charged ions may arise by other than the mechanism

$$m_1 + e \rightarrow m_1^{++} + 3e$$

and that charge separation (autoionization) can occur:

$$m_1^+ \rightarrow m_1^{++} + e$$

For example, ions which undergo the latter transition in the first drift region are transmitted when the electrostatic analyzer voltage is reduced to 0.5 of its normal value; they appear at a position $m_1/4$ on the mass scale. Such charge separation occurs in the 1,3,5-trichlorobenzene molecular ion, as evidenced by weak signals at m/z 45, 45.5, and 46 (in the correct abundance ratios for the presence of three chlorine atoms and a 3 to 1 ratio of ^{35}Cl to ^{37}Cl), when the above operation is performed.[9]

These two refocusing procedures have relative advantages and disadvantages; e.g., the accelerating voltage scan may be employed to deduce all precursors of a given daughter ion in a single scan. In contrast, the adjustment of the electrostatic analyzer voltage must be performed separately for each metastable peak; therefore, it is more laborious in practice. On the other hand, when the electrostatic analyzer voltage is reduced, the accelerating voltage remains constant; consequently, the penetration of the source by the accelerating voltage does not change. A corollary to this point arises because maximum sensitivity is obtained by using the largest possible accelerating voltage. Consequently, reducing the electrostatic analyzer voltage, whilst maintaining the maximum possible accelerating voltage, is frequently more sensitive than scanning the accelerating voltage upward from a relatively low value.

It is also possible to adjust the instrument to detect the dissociation of metastable ions in the second field-free region, which were themselves formed by metastable transitions in the first field-free region. This *double refocusing* technique may be performed to examine the general two-step dissociation pathway:

$$m_1^+ \rightarrow m_2^+ \rightarrow m_3^+$$

The magnetic analyzer is adjusted so as to transmit m_3^+ ions that are formed by dissociation of m_1^+ ions in the second field-free region. The electrostatic analyzer voltage is then decoupled from the accelerating voltage and reduced from E_0 to $(m_2^+/m_1^+)E_0$. Thus, only m_2^+ daughter ions which have kinetic energy $(m_2^+/m_1^+)zeV$ are transmitted through the electrostatic analyzer; these ions are m_2^+ ions formed by decomposition of m_1^+ ions in the first field-free region. If these ions enter the magnetic analyzer, without undergoing any further dissociation, they are not transmitted because they have too much kinetic energy. However, if decomposition occurs in the second field-free region to give m_3^+, the m_3^+ ions are produced with a kinetic energy $(m_3^+/m_2^+) \times (m_2^+/m_1^+)zeV = (m_3^+/m_1^+)zeV$. Consequently, since the magnetic analyzer is tuned to pass ions of kinetic energy $(m_3^+/m_1^+)zeV$, they are transmitted through the magnetic analyzer and arrive eventually at the collector. This technique may be used to detect consecutive dissociations; an example is given in the previous section.[4]

Application of the various refocusing procedures described above on conventional geometry instruments (e.g., that represented in Fig. 3-2) yields information concerning (all) the precursors of a given daughter ion. However, by employing a *linked scan*[10] in which the magnetic and electric fields are varied simultaneously, such that their ratio (B/E) remains constant, it is possible to elucidate (all) daughter ions produced by dissociation of a given parent ion. Information concerning the shapes of the metastable peak corresponding to a given reaction is not accessible from linked scans; in contrast, when metastable transitions are observed using the refocusing procedures discussed previously (e.g., the HV scan), the actual shape of the associated metastable peak may be recorded. The peak shape contains valuable information about the kinetic energy release which accompanies decomposition (Sec. 3-2D).

When instruments having "reversed geometry" are used, it is possible to obtain information concerning (all) daughter ions of a given parent; moreover, the kinetic energy release data are also available. Although the following discussion does not refer to metastable decompositions in the first field-free region, it is most conveniently presented here.

The essential features of a reversed-geometry mass spectrometer are shown in Fig. 3-6; ions pass through the magnetic analyzer before entering the electrostatic analyzer. A given ion, m_1^+, produced in the source, can be transmitted exclusively through the magnetic analyzer. For normal mode operation in the absence of dissociation, the ion enters the electrostatic analyzer with a kinetic energy zeV; since the electric field is adjusted to transmit ions of kinetic energy zeV, such an ion passes through the electrostatic analyzer and is eventually detected at the collector. Suppose dissociation occurs in the second field-free region to give a daughter ion, m_2^+ (closed circles in Fig. 3.6), and a neutral species (open circles); the daughter ion enters the electrostatic analyzer with a kinetic energy $(m_2/m_1)zeV$. This kinetic energy is less than that appropriate for transmission through the electrostatic analyzer; however, these daughter ions can be transmitted exclusively by reducing the electrostatic analyzer voltage from the original value, E_0, to $(m_2/m_1)E_0$. Under these conditions, the daughter ion passes through to the

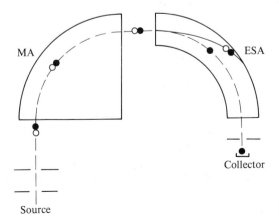

MA

ESA

Collector

Source

Figure 3-6

collector (broken curve) and is detected. In contrast, any normal m_1^+ ions, which were formed in the source and transmitted through the magnetic analyzer, have a kinetic energy zeV; consequently, they possess too much kinetic energy to pass through the electrostatic analyzer. Therefore, these ions are deflected insufficiently by the electric field and hit the outer wall of the electrostatic analyzer (solid curve). Moreover, any daughter ions m_2^+ formed in the source do not arrive at the collector because the magnetic field is adjusted to permit passage of only m_1^+ ions. Thus, by scanning E downward from E_0, all daughter ions of a given parent ion may be detected. This procedure is known as *mass-analyzed ion kinetic energy spectroscopy* (MIKES),[11] or occasionally as *direct analysis of daughter ions* (DADI).

It is possible to perform double refocusing experiments using reversed geometry instruments by transmitting metastable ions produced in the first field-free region through the magnetic analyzer. Dissociation of m_2^+ ions formed in this way to give m_3^+ ions can be detected by reducing the electrostatic analyzer voltage from E_0 to $(m_2/m_1)E_0$.

The same overall information may be acquired using conventional or reversed-geometry instruments. In some respects, this information can be obtained more conveniently by means of the MIKES technique on reversed-geometry machines. Nevertheless, the main point is that metastable transitions can be observed using either geometry, and these transitions correspond to a reaction occurring in a field-free region under relatively well-defined conditions.

C. Metastable Transitions in the Third Field-Free Region

The dissociation of an ion in the third field-free region immediately preceding the collector can be detected if the collector is equipped with a variable ion-repelling voltage.[12] Normal ions which were formed in the source and transmitted through the instrument possess a kinetic energy zeV; hence, a repeller voltage of zeV is required to prevent these ions from reaching the collector. Daughter ions, m_2^+, formed by decomposition of m_1^+ in the third field-free region have a kinetic energy $(m_2/m_1)zeV$; consequently, a smaller repeller voltage of $(m_2/m_1)zeV$ is needed to stop these ions from arriving at the collector. However, this method of detecting metastable ions has not found much practical application.

D. The Shapes of Metastable Peaks

Metastable peaks can have a variety of possible shapes; e.g., in Fig. 3-3 a roughly gaussian peak and a flat-topped peak arise in the mass spectrum of *p*-amino-anisole. These shapes are caused by the release of kinetic energy accompanying dissociation. In the discussion below, the variation of the shape of the metastable peak is considered as the nature of the associated kinetic energy release is changed. The arguments are developed for metastable ions dissociating in the second field-free region of a double-focusing instrument having conventional geometry. Exactly analogous arguments may be applied for dissociations occurring in the field-free regions of instruments with different geometries.

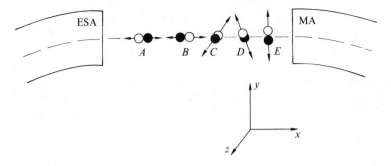

Figure 3-7

Consider the general reaction

$$m_1{}^+ \rightarrow m_2{}^+ + (m_1 - m_2)$$

occurring in the second field-free region; closed circles in Fig. 3-7 denote the incipient product ions whereas the nascent neutrals are represented by open circles. In the absence of repulsion between the neutral and fragment ions, a relatively slow separation takes place and the molecules drift apart. The resulting metastable peak is gaussian in shape and relatively narrow (Fig. 3-8*a*).

On the other hand, the fragment ion and neutral may repel each other; in this case, a release of kinetic energy accompanies dissociation. The ion and neutral receive equal and opposite momenta as a result of this "molecular explosion", in

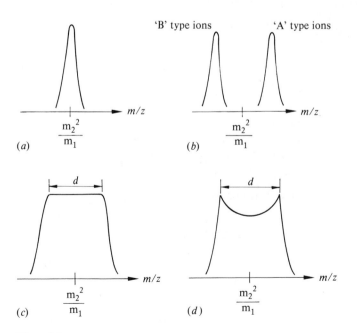

Figure 3-8

addition to that which they would have possessed in the absence of kinetic energy release. Suppose the kinetic energy release is relatively large and specific (T eV); this value is small compared to the overall kinetic energy (ca. 2–8 kV) of the ions. Nevertheless, fragment ions formed in various orientations will behave differently under the influence of the magnetic field. Decompositions occurring from orientation A (Fig. 3-7) give rise to fragment ions with slightly more momentum than would be expected in the absence of kinetic energy release. These ions are "pushed along the beam" (x direction). Consequently, they are deflected slightly less readily by the magnetic analyzer (MA) and appear at an m/z value marginally higher than the normal value (m_2^2/m_1). Conversely, dissociations in configuration B give rise to fragment ions with slightly reduced momentum because these ions are "pushed back along the beam". As a result, they are deflected somewhat more readily by the magnetic field and are recorded at an m/z value marginally lower than m_2^2/m_1 (Fig. 3-8b). Dissociation also takes place in all intermediate configurations (for example, C, D, and E); this behavior produces an "exploding sphere" of fragment ions. The net result is a flat-topped metastable peak (Fig. 3-8c). A mathematical treatment of the problem leads to a simple relationship between the kinetic energy release and the width (d, in units of m/z) of the flat-topped metastable peak,[13,14]

$$T = \frac{m_1^2 d^2 V}{16 m_2^3 m_3} \qquad eV \tag{3-6}$$

In some cases, there may be discrimination against fragment ions formed from decompositions in orientations such as C. This discrimination arises because of the finite length of the slits, which are aligned in length along the z direction. Some ions are deflected so extensively, as a result of the kinetic energy release, that they are not able to pass through the slits. The effect of this discrimination is to "hollow out" the center of the peak which becomes dish-topped in appearance (Fig. 3-8d).[14]

Different formulas relate the width of the metastable peak to the kinetic energy release when the ions are observed using refocusing procedures.[15] For a high-voltage scan,

$$T = \frac{m_2^2 V}{16 m_1 m_3} \times \left(\frac{\Delta V}{V}\right)^2 \qquad eV \tag{3-7}$$

where V is the accelerating voltage required to transmit the daughter ions formed in the first field-free region and ΔV is the width of the metastable peak expressed in volts. Similarly, for an IKE or MIKE spectrum,

$$T = \frac{m_1^2 V}{16 m_2 m_3} \times \left(\frac{\Delta E}{E}\right)^2 \qquad eV \tag{3-8}$$

where V is the accelerating voltage, E is the electrostatic analyzer voltage required to transmit daughter ions formed in the second field-free region, and ΔE is the width of the metastable peak expressed in volts.

The fundamental point is that a flat-topped or dish-topped metastable peak indicates that the dissociation is accompanied by a relatively large and specific kinetic energy release. In contrast, a gaussian peak reveals that decomposition does not involve the release of a large and discrete amount of kinetic energy. However, a value for the average kinetic energy release may be deduced from the width of a gaussian metastable peak.[16]

It is possible that a relatively small (and possibly somewhat indiscrete) kinetic energy release may give rise to a metastable peak which is not flat-topped when a comparatively large excess energy is present in the transition state for the dissociation step. This excess energy tends to broaden the peak (Sec. 4-3).

For gaussian metastable peaks, the average kinetic energy release is very sensitive to the excess energy which is present in the transition state for the final step.[15] As the quantity of excess energy increases, so the average kinetic energy release becomes larger.

Occasionally, a composite metastable peak is observed. This indicates that two, or more, competing channels exist for the dissociation process.

Metastable peaks are in effect a direct picture of what happens when dissociation takes place. The occurrence of repulsion between the ion and neutral produced may be easily detected and measured. Thus, any translational portion of a reverse activation energy, which corresponds to collisional energy being required to cause the reverse reaction, is evidenced by the appearance of a flat-topped (or dish-topped) metastable peak. These data are inaccessible to solution experiments because collisions between molecules conceal the distinction between translational and internal components of any reverse activation energy. Information about chemical reactions which is available from metastable peak shapes is considered in Secs. 4-1B, 4-2, and 4-3.

3-3 REACTION RATES AND THE MASS SPECTRUM

A. Fundamental Assumptions

The mass spectrum arises by decomposition of an assemblage of molecular ions under conditions of kinetic control. The individual ions in this assemblage do not interact because the pressure in the mass spectrometer is so low (ca. 10^{-6} torr) that collisions between the ions are effectively precluded. Consequently, the internal energies of the ions remain in a nonequilibrium distribution from the instant of ionization. This is in stark contrast with the common kinetic situation, encountered in solution, where molecules are continually exchanging energy by collisions and a Maxwell-Boltzmann distribution of energies is established. The Maxwell-Boltzmann distribution may be defined by a temperature; however, this familiar solution parameter is essentially meaningless for isolated ions in the mass spectrometer.

In order to calculate the extent of competition between two or more reactions, it is necessary to calculate the relative values of the unimolecular rate constants (k) for a given internal energy (E). This requirement arises because there is a distribution of internal energies in the parent ions (Sec. 2-1B). Several assumptions must be made before these calculations can be attempted. The most important of these assumptions are:

1. The unimolecular processes leading to the formation of the ions in the mass spectrum consist of a series of competing, consecutive unimolecular reactions of energetically excited parent ions.
2. The available internal energy randomizes throughout the molecule at a rate which is fast compared to the rate of dissociation.

In other words, although a certain amount of energy may be localized in a specific bond at the moment of ion formation, this energy may flow into other bonds in times comparable to molecular vibrational periods. Decomposition only occurs when sufficient energy accumulates in the necessary bonds.

These assumptions are reasonable and form the basis of the *quasi-equilibrium theory* (QET). According to the original and simplest form of the quasi-equilibrium theory,[17] the relationship between the rate constant k for a given reaction and the internal energy E of the ion may be expressed as follows:

$$k = v\left(\frac{E - E_0}{E}\right)^{s-1} \tag{3-9}$$

where E_0 is the activation energy for the reaction in question, v may be regarded as a frequency factor, and s is the effective number of oscillators. Consideration of the physical significance of v and s shows that Eq. (3-9) is an intuitively reasonable relationship. If the internal energy of the ion is very much larger than the activation energy, then $(E - E_0)/E$ tends to unity and the rate constant approaches the frequency factor. So for a simple single-bond cleavage reaction in the hypothetical case where the internal energy is extremely large, reaction will occur in one vibration and $k \simeq v \simeq$ vibrational frequency of the bond in question (10^{13}–$10^{14}\,\text{s}^{-1}$). For the sake of simplicity, v is regarded as a constant for any particular reaction and for a simple bond cleavage v may be approximated by the bond vibrational frequency.

If E is greater than the activation energy E_0, though of the same order of magnitude, then reaction does not occur in one vibration, but only when the requisite number of vibrational quanta (equal to the activation energy) is gathered in the reaction coordinate. The time taken for this will depend on the total number of vibrational quanta available (equivalent to an energy E in the ground electronic state of the ion), the activation energy (E_0), and the molecular size. For larger molecules, there are fewer vibrational quanta per bond for a given E; consequently, fluctuation of E_0 into the reaction coordinate is less probable and hence k is smaller. Equation (3-9) allows for this because $(E - E_0)/E$ is always less than unity; therefore, a large value of s, which is proportional to the number of atoms in the molecule, results in a small value of k. In the derivation of Eq. (3-9), s is the number of oscillators (i.e., vibrational modes) in the molecule. For nonlinear molecules, s is given by $3n - 6$, where n is the number of atoms in the molecule. However, the derivation of Eq. (3-9) involves numerous approximations and it is found in practice that for molecules investigated so far s must be taken as the effective number of oscillators. Typically, s may be considered to be one-half or one-third of the total number of oscillators, that is, s is given by $(3n-6)/2$ or $(3n-6)/3$. For energies very near the threshold, where a very rapid rise of k with E is usually observed, the effective number of oscillators which is most appropriate may be as little as one-fifth of the total.

A different situation occurs for rearrangement reactions, such as elimination of HCN (or HNC) from ionized benzonitrile (**3**) or for CH_2O loss from ionized anisole (**4**). The accumulation of the energy of activation in the appropriate coordinates is no longer a sufficient condition to cause reaction. In addition to this requirement, there is a geometrical restraint on the conformation of the transition state. For example, in the two rearrangements referred to above, the hydrogen atom which migrates must take up a specific orientation with respect to the atom to which the new bond is formed.

The relatively low probability of attaining such a specific orientation reduces the reaction rate; therefore, the rates of these rearrangement reactions do not approach bond vibrational frequencies, even at extremely high internal energies. Instead, v is lowered by some characteristic factor, and this factor increases as the geometrical restraint associated with the rearrangement becomes more stringent. This situation is analogous to the negative entropies of activation which apply to the formation of highly ordered transition states in conventional solution kinetics.

B. Competing Reactions

Three k versus E curves are reproduced in Fig. 3-9. The curves apply to a 12-atom molecule; hence, $3n-6=30$ and one-third of the total number of oscillators is taken to give the effective number of oscillators s, so that $s=10$. The parameters used for curve A are $v=10^{13}\,s^{-1}$, $E_0=2\,eV$; for curve B, $v=10^9\,s^{-1}$, $E_0=1\,eV$; and for curve C, $v=10^9\,s^{-1}$, $E_0=2\,eV$. These parameters correspond to reaction A being a simple bond cleavage whereas B and C are rearrangement processes. The lower activation energy for the rearrangement B, compared to the simple bond cleavage A, is quite realistic because rearrangement processes often involve transition states where new bonds are being formed whilst other bonds are being broken.

Suppose that the reactions corresponding to curves A, B, and C are competing unimolecular reactions occurring from a given molecular ion. It is apparent that at any given value of the internal energy, E, sufficient to cause decomposition on the mass spectrometer time scale, reaction C is several orders of magnitude slower

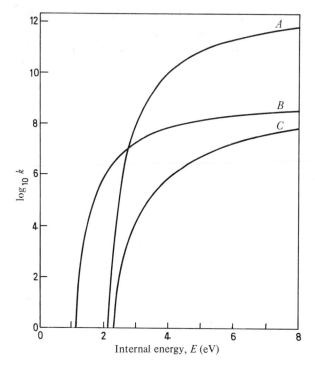

Figure 3-9

than reaction A; it is also much slower than reaction B, except at the highest internal energies. Therefore, the abundance of the metastable ion ($\log_{10} k \simeq 4\text{–}6$) and of the daughter ion ($\log_{10} k \gg 6$) resulting from reaction C will be negligible. Indeed, the existence of reaction C is purely hypothetical since it is effectively unable to compete with reactions A and B.

Ions which possess internal energies of about 1.5–2.3 eV undergo reaction B with rate constants in the range $\log_{10} k = 4\text{–}6$; no other reaction can occur at a significant rate (Fig. 3-9). Thus, molecular ions with these internal energies give rise to a metastable peak for reaction B. It is evident that these ions are able to undergo roughly 10^8 vibrations before decomposing in metastable transitions. The occurrence of reaction A at the same range of rate constant ($\log_{10} k = 4\text{–}6$) is possible only for ions with a much narrower band of internal energies (about 2.3–2.5 eV; see Fig. 3-9). This is because reaction A has a larger frequency factor than reaction B; consequently, the rise of k with E is much more rapid than is the case for reaction B.[18] On these grounds alone, a much less abundant metastable peak is expected for reaction A.[19] A second consideration is still more important: at internal energies of 2.3–2.5 eV, reaction A occurs on average much more slowly than reaction B. Therefore, the metastable transitions consist almost exclusively of reaction B; the metastable peak due to reaction A is either very low or negligible in abundance.

At higher internal energies in excess of about 2.7 eV, reaction A dominates because it occurs faster than reaction B. As a result, the daughter ion produced by reaction A is more abundant than that formed by reaction B at these higher internal energies. The relative abundances of these daughter ions correspond to the integral of the associated rate constants for their formation over the entire energy distribution. Hence, the observation that a daughter ion due to one reaction is more abundant than that produced by another competing reaction does not necessarily imply that the first reaction always occurs faster than the second. It may proceed more slowly at certain internal energies and faster at others. A useful generalization for competing reactions is that the relative activation energies are most important in determining relative ion abundances near the threshold, i.e., in low-electronvolt spectra and metastable transitions. The most abundant metastable peak corresponds to the lowest activation energy process. In contrast, for ions of high internal energies, relative reaction rates are largely influenced by the relative values of the frequency factors (Fig. 3-9). In mass spectra, it is often found that ions arising from simple bond cleavages (large v) increase in relative abundance as the electron beam energy is increased. Conversely, rearrangement reactions are often dominant in low-voltage (e.g., 10–12 eV) spectra.[20] This is because bond formation, concurrent with bond breaking in the transition state, usually lowers activation energies relative to those found in simple cleavage processes. From another viewpoint, it is clear that reactions with low-frequency factors can only compete successfully if they have lower activation energies than alternative processes which have high-frequency factors. Thus, rearrangement processes, when observed to a significant extent, must have lower activation energies than any alternative single-bond cleavage reactions.

$$C_6H_5CH_2COCH_3 \rbrack^{\ddagger}_5 \quad \xrightarrow{-CH_2=C=O} \; C_7H_8^{\ddagger} \; m/z\,92$$
$$\xrightarrow{-CH_3\overset{.}{C}=O} \; C_7H_7^+ \; m/z\,91$$

$$C_6H_5COOCH_2CH_2CH_2CH_3 \rbrack^{\ddagger}_6 \quad \xrightarrow{-C_4H_7\cdot} \; C_6H_5C\overset{\overset{+}{OH}}{\diagdown}_{OH} \; m/z\,123$$
$$\xrightarrow{-C_4H_9O\cdot} \; C_6H_5\overset{+}{C}=O \; m/z\,105$$

$$C_6H_5CH_2CH_2OH \rbrack^{\ddagger}_7 \quad \xrightarrow{-CH_2=O} \; C_7H_8^{\ddagger} \; m/z\,92$$
$$\xrightarrow{-\cdot CH_2OH} \; C_7H_7^+ \; m/z\,91$$

$$C_6H_5CH_2(CH_2)_8CH_3 \rbrack^{\ddagger}_8 \quad \xrightarrow{-C_9H_{18}} \; C_7H_8^{\ddagger} \; m/z\,92$$
$$\xrightarrow{-C_9H_{19}\cdot} \; C_7H_7^+ \; m/z\,91$$

Table 3-2 Relative metastable and daughter ion abundances in some simple cleavage and rearrangement reactions

Molecular ion	Daughter ion	Daughter ion, %	$\dfrac{[\text{Metastable}]}{[\text{Daughter ion}]}$, %
5	$m/z\,92$ (rearr.)	25	3.2
	$m/z\,91$	86	<0.01
6	$m/z\,123$ (rearr.)	57	0.18
	$m/z\,105$	100	0.003
7	$m/z\,92$ (rearr.)	60	0.58
	$m/z\,91$	100	<0.01
8	$m/z\,92$ (rearr.)	100	0.45
	$m/z\,91$	82	<0.001

It is quite common for k versus E curves for competing unimolecular reactions from a given molecular ion to cross. For instance, the following four molecular ions (**5** to **8**) undergo the reactions shown to give abundant daughter ions. In each case, one reaction is a simple bond cleavage and the other is a rearrangement process.

In Table 3-2 the abundance of each of the daughter ions in 75 eV spectra is given (as a percentage relative to the base peak), as is also the ratio [metastable]/[daughter] (percent) for each reaction.[19]

For **5**, **6**, and **7**, the simple bond cleavages give rise to the more abundant daughter ion, but the metastable ion abundances for these reactions are very low (as also for simple cleavage in **8**). In all cases, it appears that the rearrangement reaction has the lower activation energy and so gives rise to the more abundant metastable ion. The simple cleavage reactions, with their associated higher frequency factors, occur more predominantly at higher internal energies. Consequently, metastable abundances are not simply proportional to daughter ion

Table 3-3 Relative abundances of daughter ions produced by simple cleavage and rearrangement reactions at low voltage (10–16 eV) and at 70 eV

Compound	Processes	Abundance ratio†	
		Low eV	70 eV
$n\text{-}C_4H_9COC_6H_5$	$M^+ - C_3H_6 : C_6H_5CO^+$	10:1.0	1.0:2.0
$CH_3CO(CH_2)_4C_6H_5$	$M^+ - H_2O : C_7H_7{}^+$	2.0:1.0	1.0:13
$CH_3CH_2CH_2CH_2C_6H_5$	$M^+ - C_3H_6 : M^+ - C_3H_7$	13:1.0	1.0:2.0
$CH_3COCH_2COOC_2H_5$	$M^+ - C_2H_5OH : M^+ - OC_2H_5$	22:1.0	1.0:2.0

† In the quoted abundance ratios, the value for the less-abundant ion is arbitrarily taken as 1.0 unit.

abundances but, where competing reactions are involved, are determined primarily by the relative activation energies.

A comparison of relative daughter ion abundances due to some simple cleavage processes and rearrangement reactions at low voltage (10–16 eV) and at 70 eV is given in Table 3-3.[20] The lower activation energy processes dominate at low eV, but the high frequency factor processes dominate at 70 eV.

The above kinetic properties of rearrangements may also be diagnostically very useful in identifying reactions which may be formulated as direct cleavages, but which in fact involve anchimeric assistance by another group. For example, in the spectra of meta- and para-substituted β-phenylethyl bromides (9), the abundance of $M^+ - Br$ increases, relative to that of the $M^+ - CH_2Br$ ion, as the electron beam energy is lowered.[21] These results implicate a lower frequency factor for the elimination of a bromine atom and hence a more ordered transition state, probably involving aryl participation (10).

C. Factors Affecting Ion Abundances

The relative abundances of molecular, metastable, and daughter ions for a given reaction ought to be determined to a large extent by the shape of the k versus E curve for the process concerned. For the purposes of illustration, it is convenient to make the approximation that a specific rate constant, k, is sufficient to define an ion as a molecular, metastable, or daughter ion. This ignores the fact that a specific k does not correspond to a single ion lifetime; rather, all ions with a specific k are assumed to possess the mean lifetime associated with that particular rate constant. Consider a single reaction, $M^+ \rightarrow A^+$, which has the k versus E curve depicted in Fig. 3-10. The relative portions of the energy distribution A which give rise to molecular, metastable, and daughter ions may be assigned as $(M_0^+ + M_e^+)$, m*, and A^+, respectively (Fig. 3-10). If the assumption that a specific k corresponds to a single ion lifetime is not made, a more sophisticated analysis which allows for the exponential decay of ions with a given k must be employed. However, this treatment merely results in some overlapping of the areas of the energy distribution which give rise to molecular, metastable, and daughter ions (Fig. 3-11).

The most important factors influencing normal and metastable ion abundances may now be illustrated with reference to Fig. 3-11. The molecular ion abundance $(M_0^+ + M_e^+)$ corresponds largely to the area under the $P(E)$ versus E curve between the ionization energy and the appearance energy of A^+. Hence, the overriding importance of the difference $AE - IE$ in determining the molecular ion abundance becomes apparent. Given similar shapes of energy distributions,

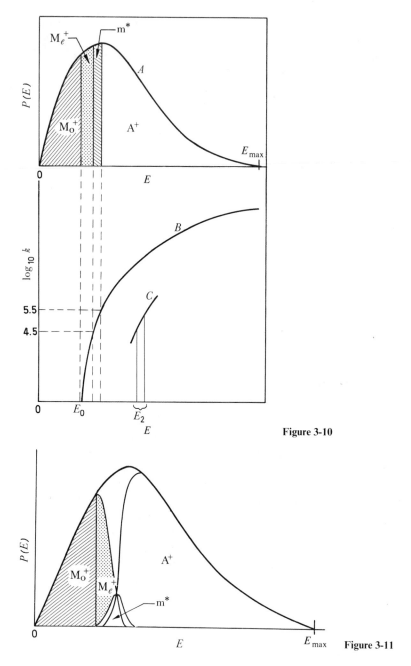

Figure 3-10

Figure 3-11

AE − IE should be a dominant factor in determining molecular ion abundances for different organic compounds at similar electron beam energies. Table 3-4 illustrates this trend in some 20 eV mass spectra (ΣD^+ represents the total daughter ion abundances. The fact that competing consecutive reactions occur from the

Table 3-4 Molecular ion abundances in relation to activation energy

Compound	IE	Lowest AE	AE − IE	$[M^+]/[\Sigma D^+]$
$p\text{-}NO_2 \cdot C_6H_4OCH_2C_6H_5$	9.1	10.0	0.9	0.03
$C_6H_5OCH_2C_6H_5$	8.4	9.7	1.3	0.28
$p\text{-}NH_2 \cdot C_6H_4OCH_2C_6H_5$	7.6	9.7	2.1	0.33
$CH_3(CH_2)_3CO_2Me$	10.4	10.9	0.5	0.007
$p\text{-}MeO \cdot C_6H_4 \cdot (CH_2)_3CO_2Me$	8.0	10.6	2.6	0.17

molecular ion in no way invalidates the illustration, and the lowest appearance energy is employed in each case.

Since the initial rise of the rate constant with internal energy is rapid, the segments of the energy distribution which yield $M_e{}^+$ and m* are relatively narrow. The area of these segments will be increased by factors which decrease the slope of the curve B (Fig. 3-10) between $\log_{10}k = 3$ and 6. That is, in general, an increase in molecular size or a decrease in the frequency factor v should tend to increase the relative abundances of $M_e{}^+$ and m*.

The most abundant metastable peak for decomposition of a given ion generally corresponds to the decay route with the lowest activation energy (see Sec. 3-3B). Higher activation energy processes, e.g., a reaction having a k versus E curve such as C (Fig. 3-10), do not usually give rise to metastable peaks. This is because ions in the energy segment E_2, which could give rise to dissociation at the appropriate rate, decompose preferentially by reaction B at a much faster rate. Thus, ions of energy E_2 would have decomposed long before reaching the field-free regions.

The situation regarding the formation of A^+ from higher energy molecular ions is normally not as straightforward as depicted in Fig. 3-10. Competition may occur with other reactions of M^+ and the relative abundances of the associated daughter ions would then depend on the relative k versus E curves for the various processes (see Sec. 3-3B). In addition, molecular ions with more than a certain internal energy can yield primary daughter ions which themselves have enough energy to decompose to form secondary daughter ions (for example, $M^+ \rightarrow A^+ \rightarrow C^+$). As the electron beam energy is increased, the energy distribution contains more ions of higher internal energies; consequently, there is a corresponding increase in the abundance of secondary daughter ions. The overall ion abundances of A^+ and C^+ (and the corresponding metastable ions) are determined not only by the $P(E)$ and $k(E)$ curves for the molecular ion but also by the analogous parameters for A^+. The energy distribution of A^+ formed in the source is influenced by the energy carried off by the neutral fragment lost when M^+ decomposes into A^+.

The differences between high- and low-voltage spectra may also be understood in terms of Fig. 3-10. As the electron beam energy is decreased the energy distribution contains more molecular ions with lower internal energies. Therefore, ions formed by reactions which require more energy diminish in abundance

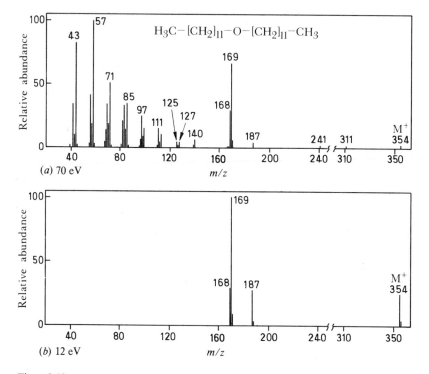

Figure 3-12

relative to the abundance of the molecular ion, or disappear altogether from the mass spectrum. This conclusion applies regardless of whether the relevant daughter ions are produced from the molecular ion in one or more steps. Consequently, low-electronvolt spectra are much simpler than 70 eV spectra. This is illustrated in Fig. 3-12, where the mass spectra of di-*n*-dodecylether at 70 eV and 12 eV are compared.[22] In the 12 eV spectrum, only the ions of lowest appearance energy, generated by the reactions shown below, are evident.

$$C_{12}H_{25}OC_{12}H_{25} \overset{\ddagger}{} \begin{array}{l} \xrightarrow{-C_{12}H_{23}\cdot} C_{12}H_{25}\overset{+}{O}H_2 \xrightarrow{-H_2O} C_{12}H_{25}{}^+ \\ m/z\,187 m/z\,169 \\ \xrightarrow{-C_{12}H_{25}OH} C_{12}H_{24}{}^+ \\ m/z\,168 \end{array}$$

It is clear that relative appearance energies are an important factor in determining which fragmentations will predominate in bifunctional or poly-functional compounds. Provided that the substituents do not interact directly through space in the decomposition reactions or strongly through bonds (e.g.,

Table 3-5 Appearance energies for fragmentation of some C_6H_5X compounds

X	Neutral fragment	AE, eV
$COCH_3$	CH_3	9.99
$C(CH_3)_3$	CH_3	10.26
$CH(CH_3)_2$	CH_3	10.65
CO_2CH_3	OCH_3	10.80
$N(CH_3)_2$	H	10.80
CHO	H	10.99
C_2H_5	CH_3	11.25
OCH_3	CH_2O	11.30
I	I	11.46
OH	CO	11.67
CH_3	H	11.80
Br	Br	12.02
NO_2	NO_2	12.16
NH_2	HCN	12.50
Cl	Cl	13.20
CN	HCN	14.60
F	C_2H_2	14.73

resonance interaction), the predominant fragment should be predictable from the appearance energy of the *characteristic fragment ion* of each substituent. Thus, for example, in order to provide a reference series for the *relative ease of fragmentation* of different groups in disubstituted benzenes, the appearance energies have been measured[23] for the characteristic fragmentations of X from monosubstituted benzenes $C_6H_5X^+$. The order of appearance energies is shown in Table 3-5.

Since the reactions of lowest appearance energy in the series of monosubstituted benzenes (Table 3-5) also have high frequency factors (for example, $CH_3\cdot$ loss from $C_6H_5COCH_3^+$, $CH_3\cdot$ loss from $C_6H_5C(CH_3)_3^+$, $OCH_3\cdot$ loss from $C_6H_5COOCH_3^+$), these reactions will *always* occur faster than, say, CH_2O loss from $C_6H_5OCH_3^+$, $Cl\cdot$ loss from $C_6H_5Cl^+$, CO loss from $C_6H_5OH^+$, or HCN loss from $C_6H_5CN^+$, when two such very different groups are attached to the same benzene ring. It is found that the above generalization holds when the substituents are chosen with very different activation energies (see, for example, the 18 eV spectrum of *p*-cyano-*t*-butylbenzene in Fig. 3-13). Even the difference of 0.48 eV in AE for fragmentation of Br and NH_2 substituents appears still to be reflected in the 18 eV spectrum of *p*-bromoaniline (Fig. 3-14). However, when substituents fall as close as this to each other in Table 3-5, modification of the relative activation energies must be considered because competition occurs between loss of $Cl\cdot$ and HCN in the 18 eV spectrum of *p*-chloroaniline (Fig. 3-15).

No predictions can be made concerning the mass spectra of ortho-disubstituted benzenes since the activation energies for decomposition may be drastically changed by direct interaction of the two substituents. For para- and meta-disubstituted benzenes and other aromatic disubstituted systems where the two

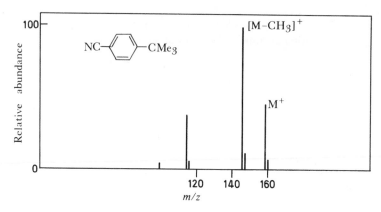

Figure 3-13

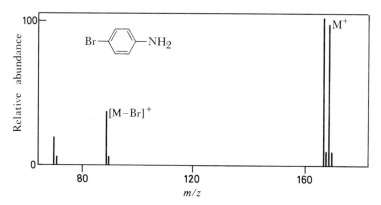

Figure 3-14

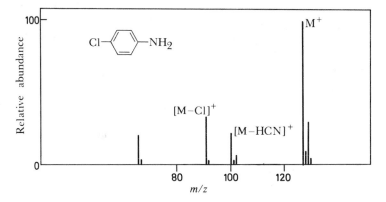

Figure 3-15

substituents cannot interact directly through space, Table 3-5 constitutes a useful "fragmentation probability" table.

Such results show that:

1. The appearance energies for decomposition in the monosubstituted benzenes are not *drastically* altered in meta- and para-disubstituted benzenes; nevertheless, they are slightly modified by interaction between the two substituents. Therefore, the mass spectrum of a disubstituted benzene cannot confidently be predicted from the relative order given in Table 3-5 if the two substituents fall close together (for example, CH_3 and Br).
2. The dissociations involving the two substituents are truly competing, at least in most cases, for internal energies at which fragmentation of both substituents is, in principle, possible. In this respect, the spectra reproduced in Figs. 3-13, 3-14, and 3-15 support the basic tenets of the quasi-equilibrium theory.

3-4 FIELD IONIZATION KINETICS

The technique of field ionization kinetics (FIK) was devised using a single-focusing mass spectrometer.[24] It was subsequently used, to advantage, with double-focusing mass spectrometers.[25] It is the only technique which enables the kinetics of gas-phase ion decompositions to be measured in the 10^{-9}–10^{-12} s time region.

A schematic illustration of the variation in potential within a field ionization source is given in Fig. 3-16. A blade or wire is at $+8\,kV$ relative to a closely located cathode slit (Sec. 1-2C). The potential again rises as the ion beam passes through focus electrodes, but the ions finally have a translational energy of 8000 eV as they leave the source exit slit and enter the electric sector.

Suppose that a molecular ion m_1^+ decomposes to a fragment ion m_2^+ where the potential between the wire and cathode has fallen to 6000 V. At the instant of decomposition, the translational energy of m_1^+ is 2000 eV, which is partitioned

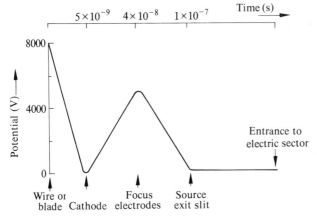

Figure 3-16

between m_2^+ and the neutral fragment according to their mass ratio. Thus, the translational energy of m_2^+ at this point is $2000 \, m_2/m_1$, and of m_2^+ as it enters the electric sector $(2000 \, m_2/m_1 + 6000) \, eV$. Hence, this particular m_2^+ ion will only be transmitted by the electric sector (set to pass ions of 8 kV energy) if the emitter potential of 8000 V is raised to compensate for the energy lost to the neutral. It is evident that ions formed by decomposition at any point between the emitter and cathode can be selected at will. Moreover, if the variation of potential with distance is known, the time of ion decomposition can be derived. An appropriate time scale has been added to Fig. 3-16.

It transpires that for normal FI and FD work, i.e., spectra obtained at constant emitter potential, fragment ions will not be transmitted if they are formed in fragmentations occurring in greater than ca. 10^{-11} s. By use of FIK, it has been shown that the molecular ion of $3,3,6,6-d_4$-cyclohexene loses predominantly C_2H_4 some 10^{-11} s after ionization, as expected on the basis of the retro-Diels-Alder reaction indicated:

At these very short lifetimes, rearrangement processes in the molecular ion complete poorly with the fragmentation. However, after only 10^{-10} s, losses of $C_2H_2D_2$ and C_2H_3D are dominant over C_2H_4 loss,[25] establishing that ionized cyclohexene can undergo an isomerization process which requires less energy that the retro-Diels-Alder reaction.

In connection with the selection of different mass spectrometric techniques to sample different ion lifetimes, it is appropriate to mention the position of field ionization metastables in the general spectrum of lifetimes. For the reaction $m_1^+ \rightarrow m_2^+$, "fast" metastables $(\log_{10} k = 8-12)$ formed by decomposition during acceleration will appear in a single-focusing mass spectrometer as a broad continuum between $m/z = m_2$ and m_2^2/m_1; and "slow" metastables $(\log_{10} k = 6-8)$ will appear at $m/z = m_2^2/m_1$, as for metastables in the EI spectrum (see also Fig. 4-11).

REFERENCES

1. R. D. Craig, B. N. Green, and J. D. Waldron: *Chimia (Aarau)*, vol. 17, p. 33, 1963.
2. I. Howe and D. H. Williams: *Chem. Comm.*, p. 220, 1968.
3. J. A. Hipple and E. U. Condon: *Phys. Rev.*, vol. 69, p. 347, 1946.
4. K. R. Jennings: *Chem. Comm.*, p. 283, 1966.
5. W. Higgins and K. R. Jennings: *Chem. Comm.*, p. 99, 1965.

6. M. Barber and R. M. Elliott: *Twelfth Annual Conference on Mass Spectrometry and Allied Topics,* Committee E.14, ASTM, Montreal, June, 1964.
7. K. R. Jennings: In R. Bonnet and J. G. Davies (eds.), "Some Newer Physical Methods in Structural Chemistry," pp. 105–109, United Trade Press, London, 1967.
8. F. W. McLafferty, J. Okamoto, H. Tsuyama, Y. Nakajima, T. Noda, and H. W. Major: *Org. Mass. Spec.,* vol. 2, p. 751, 1969.
9. J. Seibl: *Org. Mass Spec.,* vol. 2, p. 1033, 1969.
10. A. F. Weston, K. R. Jennings, S. Evans, and R. M. Elliott: *Int. J. Mass Spec. Ion Phys.,* vol. 20, p. 317, 1976.
11. J. H. Beynon, R. G. Cooks, J. W. Amy, W. E. Baitinger, and T. Y. Ridley: *Anal. Chem.,* vol. 45, p. 1023A, 1973.
12. N. R. Daly, A. McCormick, and R. E. Powell: *Rev. Sci. Instrum.,* vol. 39, p. 1163, 1968.
13. J. H. Beynon, R. A. Saunders, and A. E. Williams: *Zeitschrift für Naturforschg.,* vol. 20a, p. 180, 1965.
14. J. H. Beynon and A. E. Fontaine: *Zeitschrift für Naturforschg.,* vol. 22a, p. 334, 1967, and references cited therein.
15. R. G. Cooks, J. H. Beynon, R. M. Caprioli, and G. R. Lester: "Metastable Ions," Elsevier, Amsterdam, 1973.
16. D. T. Terwilliger, J. H. Beynon, and R. G. Cooks: *Proc. Roy. Soc.,* vol. 341A, p. 135, 1974.
17. H. M. Rosenstock, M. B. Wallenstein, A. L. Wahrhaftig, and H. Eyring: *Proc. Natl. Acad. Sci. U.S.A.,* vol. 38, p. 667, 1952.
18. W. A. Chupka: *J. Chem. Phys.,* vol. 30, p. 191, 1959.
19. F. W. McLafferty and R. B. Fairweather: *J. Am. Chem. Soc.,* vol. 90, p. 5915, 1968.
20. D. H. Williams and R. G. Cooks: *Chem. Comm.,* p. 663, 1968.
21. R. H. Shapiro and T. F. Jenkins: *Org. Mass Spec.,* vol. 2, p. 771, 1969.
22. M. Spiteller-Friedmann and G. Spiteller: *Chem. Ber.,* vol. 100, p. 79, 1967.
23. I. Howe and D. H. Williams: *J. Am. Chem. Soc.,* vol. 91, p. 7137, 1969.
24. H. D. Beckey, H. Key, K. Levsen, and G. Tenschert: *Int. J. Mass Spec. Ion Phys.,* vol. 2, p. 101, 1969.
25. P. J. Derrick, A. M. Falick, and A. L. Burlingame: *J. Am. Chem. Soc.,* vol. 94, p. 6794, 1972; A. M. Falick, P. J. Derrick, and A. L. Burlingame: *Int. J. Mass Spec. Ion Phys.,* vol. 12, p. 101, 1973.

FOUR

ENERGETICS OF UNIMOLECULAR REACTIONS

4-1 THERMOCHEMICAL DATA

A. Negligible Reverse Activation Energies

It was seen (Sec. 3-3) that the ability of one reaction to compete with another is, for slow reactions (metastable peaks), primarily determined by the relative activation energies (E_f) for the two reactions. Many single-bond cleavages do not involve a significant reverse activation energy (E_r). Energy fluctuates into the reaction coordinate until that bond finally breaks, and the products separate without a fall in potential energy (Fig. 4-1a). This situation contrasts with that (Fig. 4-1b) where the products are stabilized relative to the transition state (TS) for reaction and there is both a forward (E_f) and a reverse (E_r) activation energy. Broadly speaking, the potential energy profile shown in Fig. 4-1b corresponds to rearrangement reactions where there is largely bond breaking up to the transition state and mainly bond making after the transition state has been passed.

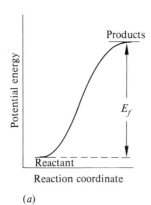

(a)   (b)   **Figure 4-1**

It follows that, as a useful approximation, the energy requirement for a single-bond cleavage is given by the difference in stabilities of reactant and products, i.e.,

$$E_f \simeq \Sigma \, \Delta H_f \, (\text{products}) - \Delta H_f \, (\text{reactant}) \tag{4-1}$$

Table 4-1 Heats of formation (ΔH_f, kJ mol^{-1}) of some ions

Ion	ΔH_f	Ion	ΔH_f
H^+	1530	$CH_4^{\ddagger}$	1150
CH_3^+	1090	$C_2H_6^{\ddagger}$	1040
$C_2H_5^+$	920	$C_3H_8^{\ddagger}$	950
$CH_3CH_2CH_2^+$	870	$C_2H_4^{\ddagger}$	1070
$(CH_3)_2CH^+$	800	$CH_3CH{=}CH_2^{\ddagger}$	960
$CH_3(CH_2)_2CH_2^+$	840	$HC{\equiv}CH^{\ddagger}$	1330
$(CH_3)_2CHCH_2^+$	830	$CH_3C{\equiv}CH^+$	1180
$CH_3CH_2\overset{+}{C}HCH_3$	770	$CH_2{=}C{=}CH_2^{\ddagger}$	1120
$(CH_3)_3C^+$	690	$H_2O^{\ddagger}$	980
		H_3O^+	580
$CH_2{=}\overset{+}{C}H$	1110	$H\overset{+}{C}{=}O$	810
$CH_2{=}CH{-}\overset{+}{C}H_2$	950	$CH_3\overset{+}{C}{=}O$	630
 (allyl/propenyl cation)	850	$CH_3CH_2\overset{+}{C}{=}O$	600
 (methallyl cation)	880	$C_6H_5\overset{+}{C}{=}O$	730
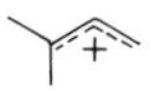 (dimethylallyl cation)	770	$CH_2{=}\overset{+}{O}H$	690
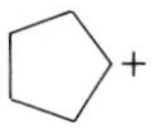 (cyclopentenyl cation)	820	$CH_3CH{=}\overset{+}{O}H$	600
$HC{\equiv}CCH_2^+$	1170	$CH_3\overset{+}{O}{=}CH_2$	640
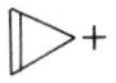 (cyclopropenyl cation)	1070	$(CH_3)_2C{=}\overset{+}{O}H$	510
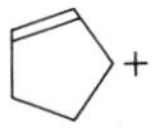 (cyclopentadienyl cation)	830	$CH_3COOH^{\ddagger}$	580
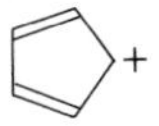 (cyclopentadienyl cation)	1070	$CH_3C(\overset{+}{O}H){=}OH$	320
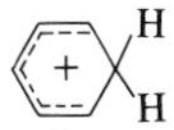 (benzenium, H H)	860		
$C_6H_5^+$	1140		
$C_6H_5CH_2^+$	890	$CH_2{=}\overset{+}{N}H_2$	740
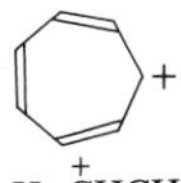 (tropylium cation)	900	$CH_3\overset{+}{N}H{=}CH_2$	640
$C_6H_5\overset{+}{C}HCH_3$	850	$CH_3CH{=}\overset{+}{N}H_2$	650
$C_6H_4^{\ddagger}$	1470	 (benzene radical cation)	970

Thus, for single-bond cleavages, if the heats of formation (ΔH_f) of the reactant and suspected products are known, it is possible to predict which of several single-bond cleavages a reactant ion will undergo. Relevant data are given in Tables 4-1 to 4-3; a much more extensive compilation of the heats of formation is available.[1]

The data in Table 4-1 give a useful guide to the relative stabilities of ionic fragments. They are largely derived by electron-impact ionization of radicals or by electron-impact ionization and fragmentation of even-electron molecules (Sec. 2-1A). Such methods are not normally suitable for obtaining ΔH_f values for protonated saturated molecules, for example, $CH_3\overset{+}{O}H_2$ and $CH_3\overset{+}{N}H_3$. These values may be obtained, however, from bimolecular studies (Chap. 5); representative data from equilibrium ICR experiments are given in Table 4-2. Molecules with a negative ΔH^0 value in Table 4-2 have a greater affinity for the proton than NH_3.

If a reaction

$$B_1H^+ + B_2 \rightleftharpoons B_1 + B_2H^+$$

proceeds to the right with an enthalpy change ΔH^0, then

$$\Delta H_f(B_1H^+) + \Delta H_f(B_2) = \Delta H_f(B_2H^+) + \Delta H_f(B_1) - \Delta H^0$$

Since the proton affinity (PA) of B is given by

$$PA(B) = \Delta H_f(B) + \Delta H_f(H^+) - \Delta H_f(BH^+)$$

then ΔH^0 will be negative [that is, B_2H^+ observed as the sole or main ionic product if $PA(B_2) > PA(B_1)$]. Thus, in an experiment where equilibrium conditions prevail [which may occasionally be the case even for unimolecular

Table 4-2 Proton affinities (ΔH^0, kJ mol^{-1}) (relative to NH_3) of some common bases (B)†

Base (B)	ΔH^0	$\Delta H_f(BH^+)$‡	Base (B)	ΔH^0	$\Delta H_f(BH^+)$‡
H_2O	133	580	$(C_2H_5)_2C{=}O$	16	447
$H_2C{=}O$	155	688	$HCONHCH_3$	~ −2	...
MeOH	84	574	$C_6H_5NH_2$	−27	750
C_6H_6	87	860	CH_3NH_2	−38	629
C_2H_5OH	65	520		−66	764
CH_3COOH	60	318			
$(CH_3)_2O$	51	455	$(C_2H_5)_3N$	−112	478
CH_3COOCH_3	29	309	$(CH_3)_2N(CH_2)_3N(CH_3)_2$	−142	...

† Values of ΔH^0 are for the equilibrium $NH_4^+ + B \rightleftharpoons BH^+ + NH_3$, and are taken from J. F. Wolf, R. H. Staley, I. Koppel, M. Taagepera, R. T. McIver, Jr., J. L. Beauchamp, and R. W. Taft: *J. Am. Chem. Soc.*, vol. 99, p. 5417, 1977.

‡ Values of ΔH_f of the corresponding protonated base are derived from the equation $\Delta H_f(BH^+) = \Delta H_f(H^+) - PA(NH_3) + \Delta H_f(B) + \Delta H^0 = 690 + \Delta H_f(B) + \Delta H^0$ kJ mol^{-1}.

dissociation at low pressures—due to ion-dipole interactions (Sec. 4-3)], a proton-bound mixed dimer $B_1\overset{+}{H}B_2$ will dissociate preferentially to B_2H^+ if $PA(B_2) >$ $PA(B_1)$. This holds even if $\Delta H_f(B_2H^+) > \Delta H_f(B_1H^+)$, as is often the case (Table 4-2).

The principles involved in deducing reaction pathways are illustrated by considering three possible single-bond cleavages which ionized ethylbenzene (**1**) might undergo:

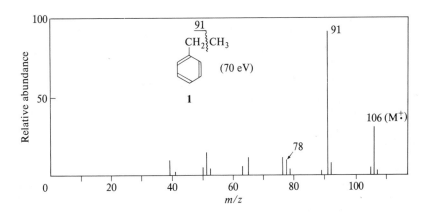

It is clear from the ΔH_f data that dissociation (4-3) should be favored over the others. Indeed, the dominant fragmentation of the ethylbenzene molecular ion is by loss of a methyl radical (Fig. 4-2). Note that since single-bond cleavages normally have high, and not very different, frequency factors, generalizations based on the energetics of these reactions may be extrapolated to faster (source) reactions, as above.

The above example does not consider the possibility that **1** isomerizes prior to dissociation, as it may if the activation energy for isomerization is less than that for

Figure 4-2

Eq. (4-3). A plausible isomerization is $1 \rightleftharpoons 2 \rightleftharpoons 3 \rightleftharpoons 4$, after which methyl radical loss could in principle occur (in competition) from **1** or **4**. However, the ΔH_f values for benzyl and tropylium (**5**) ions are rather similar (Table 4-1). Hence, in this particular case, the isomerization possibility does not change the conclusion concerning the products preferentially formed.

1, R = CH₃   **2, R = CH₃ or H**   **3, R = CH₃ or H**

6, R = H

7   **5, ΔH_f 900**   **4, R = CH₃ or H**

Consideration of the data given in Tables 4-1 and 4-3 leads to the conclusion that the toluene molecular ion (**6**) will fragment by loss of a benzylic hydrogen radical, as indeed it does. More energy is needed than for fragmentation of **1** mainly because $\Delta H_f(H)$ is larger than $\Delta H_f(CH_3)$ by $76\,kJ\,mol^{-1}$. In the toluene case, the occurrence of the reversible isomerization $6 \rightleftharpoons 2 \rightleftharpoons 3 \rightleftharpoons 4$ is indicated by loss of both H · and D · from the molecular ions generated from **7**.

It is evident that the fragmentation of an ion, if via a single-bond cleavage, will occur to give the most stable combination of ionic and neutral products. However, where rearrangement reactions are considered, which may (Fig. 4-1b) involve reverse activation energies, it can be surmised only that they are feasible if $\Delta H_f (\text{products}_R) \leqslant \Delta H_f (\text{products}_S)$, where "products$_R$" and "products$_S$" arise from fragmentations, respectively, with and without rearrangement. For example, $\Delta H_f(C_6H_6^+) + \Delta H_f(C_2H_4)$ is $\sim 1030\,kJ\,mol^{-1}$, which is approximately equal to $\Sigma \Delta H_f$ for Eq. (4-3). Thus, C_2H_4 loss from ionized ethylbenzene, which may involve a reverse activation energy, seems improbable on energetic grounds alone, and indeed it is a minor process in the 70 eV electron-impact spectrum (Fig. 4-2).

The limited data of Tables 4-1 to 4-3 can be usefully extended (in approximate form only) by a method of group equivalents.[2] Table 4-4 gives the values appropriate to the calculation of ΔH_f values of neutral molecules as isolated species. These may be used as increments for ions (Tables 4-1 and 4-2) if the substitution or insertion is remote (i.e., three or more σ bonds distant) from the cationic center. The data in Table 4-4 should not be used for substitution of a group directly attached to a cationic center. If the substitution is β to a cationic

Table 4-3 Heats of formation ($\Delta H_{f298(g)}$ kJ mol^{-1}) of some simple radicals and neutrals

Formula	ΔH_f	Formula	ΔH_f
H·	218	H_2	0
D·	222	:CH$_2$	392
CH$_3$·	142	C_2H_4	52
C$_2$H$_5$·	107	CH$_3$CH=CH$_2$	20
CH$_3$CH$_2$CH$_2$·	87	(CH$_3$)$_2$C=CH$_2$	−17
CH$_3$ĊHCH$_3$	74		
		CH$_4$	−75
CH$_2$=CH·	250	C$_2$H$_6$	−85
CH$_2$=CH—CH$_2$·	170	C$_3$H$_8$	−104
C$_6$H$_5$·	300		
C$_6$H$_5$CH$_2$·	188	C$_6$H$_6$	83
·OH	39	H$_2$O	−245
·OCH$_3$	−4	CH$_3$OH	−200
		C$_2$H$_5$OH	−235
CH$_3$Ċ=O	−23	H$_2$C=O	−117
·NH$_2$	172	CH$_3$CHO	−166
		NH$_3$	−46
·C≡N	435		
·F	79	HCN	135
·Cl	122	HF	−271
·Br	112	HCl	−92
·I	107	HBr	−36
·SH	142	HI	26
		H$_2$S	−21
CO	−110	CH$_3$COOH	−455
CO$_2$	−394	CH$_2$=C=O	−61
		SO$_2$	−297

center (i.e., at the atom adjacent to that carrying the positive charge), then the corrected values given in parentheses in Table 4-4 should be used.[3]

The use of Tables 4-1 to 4-4 is illustrated by two examples.

Example 1 The monosubstituted benzenes **8** could in principle react via loss of X or HX. The relevant ΔH_f values of the phenyl cation and ionized benzyne (Table 4-1) and those of X and HX (Table 4-3) indicate that ionized chloro-, bromo-, and iodobenzenes should lose X· and not HX, whereas fluorobenzene may lose HF unless this reaction involves a reverse activation energy $\geqslant 70$ kJ mol^{-1}.

Experimentally, it is observed that these compounds do indeed lose Cl, Br, I, and HF, both in metastable transitions and source reactions. Since in the

Table 4-4 Group equivalents for estimation of ΔH_f values (kJ mol^{-1})

Group	Increment	Group	Increment
CH$_3$—	−42	$\diagdown$C=C$\diagup$	100
—CH$_2$—	−21	—C≡C— ⎫	226
—$\overset{\mid}{C}$H—	−4	—C≡CH ⎭	
—$\overset{\mid}{\underset{\mid}{C}}$—	+4	—OH	−176 (β, −135) (γ, −165)
—CH=CH$_2$	63	—O—	−30
—CH=CH—	80	—NH$_2$	+17 (β, +60) (γ, +30)
		—Cl	−42
$\diagdown$C=CH$_2$	71		
$\diagup$		$\diagdown$C=O	−134
$\diagdown$C=C$\overset{\diagup H}{\diagdown}$	84	$\underset{O\diagup\diagdown}{\diagdown}$C—O—	−334

case of benzonitrile, $\Sigma\Delta H_f$ (products) are so close (1580 and 1600 kJ mol^{-1} for the loss of HCN and CN, respectively), then either reaction is feasible. Experimentally, HCN loss is observed in metastable transitions and both HCN and CN losses in source reactions.

8, X = F, Cl, Br, I, or CN

Example 2 Some routes which might be considered for the decomposition of ionized butyl acetate (**9**) are given in Table 4-5. The heat of formation of butyl acetate estimated from Table 4-4 is −480 kJ mol^{-1}, in close agreement with the experimental value of −490 kJ mol^{-1}.[1] The value of $\Delta H_f[\text{CH}_3\text{COO(CH}_2)_3\text{CH}_3]^{\ddagger}$, 430 kJ mol^{-1}, is derived by addition of the ionization energy (920 kJ mol^{-1}, determined by photoionization) to −490 kJ mol^{-1}. Thus, if rearrangement reactions such as (6) or (7) were to occur with little or no reverse activation energy, then the molecular ion abundance should be small or negligible. Molecular ions would require only ca.30 kJ mol^{-1} of internal energy to have the possibility of decomposing before reaching the collector.

The hypothetical reactions are split into three groups: (*a*) single-bond cleavages, (1) to (3); (*b*) fragmentations with hydrogen rearrangements, (4) to (7);

Table 4-5 Some hypothetical decomposition routes for [BuOAc]$^{+}$

Number	Products	m/z	$\Sigma \Delta H_f$, kJ mol^{-1}
1	$CH_3\overset{+}{C}{=}O + \cdot OCH_2CH_2CH_2CH_3$	43	~ 560
2	$CH_3\overset{O}{\overset{\|}{C}}{-}\overset{+}{O}{=}CH_2 + \cdot CH_2CH_2CH_3$	73	~ 530
3	$CH_3\overset{O}{\overset{\|}{C}}OCH_2CH_2 \cdot + \overset{+}{C}H_2CH_3$	29	~ 770
4	$CH_3\overset{O}{\overset{\|}{C}}{-}\overset{+}{O}{=}CHCH_3 + \cdot CH_2CH_3$	87	~ 510
5	$CH_3COOH^+ + CH_2{=}CHCH_2CH_3$	60	580
6	$CH_2{=}CHCH_2CH_3{}^+ + CH_3COOH$	56	450
7	$CH_3C\overset{\overset{+}{O}H}{\diagup}\underset{OH}{\diagdown} + CH_2{=}CH\dot{C}HCH_3$	61	470
8	$CH_3CH_2CH_2CH_2CH_3{}^+ + CO_2$	72	460
9	$CH_3O\,CH_2CH_2CH_2CH_3{}^+ + CO$	88	~ 530

and (*c*) skeletal reorganization with extrusion of extremely stable neutrals, (8) and (9). It must be concluded that among many possible single-bond cleavage reactions, (1) and (2) are the most favorable ones. This is the case in practice (Fig. 4-3*a*). If hydrogen rearrangements (6) and (7) occur with insignificant reverse activation energies, then these should be favored at low electronvolts, since the lower-energy pathways should be preferred to the higher-energy route at long ion lifetimes. Indeed, m/z 56 (Fig. 4-3*b*) becomes the base peak in 15 eV spectra. Finally, both (8) and (9) give rise to low-energy products; they are not observed in practice (Fig. 4-3). This observation is a useful general guide, i.e., reactions requiring σ-bond formation (not involving hydrogen) with elimination of stable neutrals are normally not observed if the ionic fragment to be produced does not contain sites of unsaturation (double or triple bonds); the reverse activation energies are evidently too large.

Because of the plethora of products which may in principle arise even in the mass spectrum of a simple molecule, it is usually too laborious or not feasible to approach spectral interpretation in this way. Additionally, in many cases the ΔH_f values of ions are known with insufficient accuracy to permit reliable predictions. However, Tables 4-1 to 4-4 permit a general understanding of the relative stabilities of organic ions and neutrals. Useful generalizations are that (1) production of relatively stable ions is favored, which include benzylic, allylic, and tertiary ions, and oxonium ($={\overset{+}{O}}{-}$) and imminium ($-\overset{+}{N}H{=}$) ions; (2) direct cleavage of double ($-C{\not\equiv}C-$) bonds is unfavorable and rare, as is vinylic cleavage ($-C{=}C{\not\mid}R)-$ unless there is no alternative; (3) the loss of certain stable neutrals (for example,

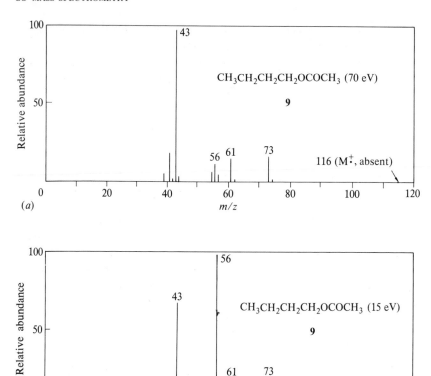

Figure 4-3

H_2O, CO, CO_2) is common, even in some cases where considerable prior re-arrangement is necessary; and (4) fragmentation to form small ionic fragments (for example, H^+, CH_3^+, H_2O^+) does not normally occur, as these small fragments have high ΔH_f values, partly due to lack of a large polarizable electron cloud which may be used to stabilize the positive charge.

B. Large Reverse Activation Energies

The ability to predict, or rationalize, how an ion will fragment in a mass spectrometer will be compromised if (1) there is no facile single-bond cleavage and (2) possible rearrangement reactions have high (but generally unknown) reverse activation energies. Reactions in the latter category include concerted, symmetry-forbidden losses of H_2.

Protonated formaldehyde, $CH_2{=}\overset{+}{O}H$, is clearly unlikely to fragment without rearrangement (prior to or during fragmentation) since the possible products, for example, $:CH_2 + \overset{+}{O}H$, are extremely unstable. Indeed, in slow reactions (metastable transitions), its sole unimolecular decomposition is by loss of H_2. Deuterium labeling (Sec. 11-2) establishes that this reaction is a 1,2-elimination:

$$H_2C=\overset{+}{O}H \rightarrow \begin{matrix} H\text{-}\text{-}\text{-}H \\ \vert \quad \vert \\ C\text{=}\text{=}\text{=}O^+ \\ \diagup \\ H \end{matrix} \quad \overset{H\text{—}H}{\rightarrow} H\text{—}C\text{≡}O^+ \qquad (4\text{-}5)$$

The reaction has a forward activation energy of $\sim 330\,\text{kJ mol}^{-1}$ and a reverse activation energy of $\sim 220\,\text{kJ mol}^{-1}$. Of this reverse activation energy, the metastable peak shape (Sec. 3-2D) establishes that $140\,\text{kJ mol}^{-1}$ is released as kinetic (translational) energy. It is interesting to enquire why this reaction has such a large reverse activation energy and why such a large fraction of this reverse activation energy appears as translation.

The reaction may be usefully analyzed in terms of the Woodward–Hoffmann rules,[4] for the reaction is a symmetry-forbidden four-electron process. The correlation diagram which is relevant to such a reaction is given in Fig. 4-4. In this type of reaction, four electrons of two $\sigma(X\text{—}H)$ bonds [left-hand side (LHS) of Fig. 4-4] become two σ electrons of H_2 and two π electrons of the other product [right-hand side (RHS) of Fig. 4-4]. In the reactant, these four electrons occupy the two molecular orbitals associated with levels 1 and 2; those associated with levels 3 and 4 are unoccupied. In comparing molecular orbitals of reactants and products, orbitals of like symmetry are connected by solid lines (A = asymmetric and S = symmetric, where the symmetry of the orbitals is assessed relative to a vertical plane passing through the midpoint of the horizontal C—X bond). It can be seen that if the electrons in the reactant orbitals simply passed into product orbitals of like symmetry, then level 3 of the products would be occupied at the

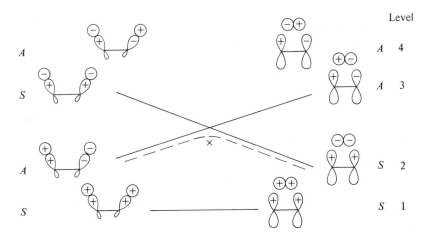

Figure 4-4

expense of level 2, i.e., the products would be produced in an electronically excited state. However, it is clear that the products of Eq. (4-5) are not produced in an electronically excited state since, after allowing for kinetic energy release ($140\,kJ\,mol^{-1}$), their internal energy is only $\sim 80\,kJ\,mol^{-1}$ above the sum of the heats of formation of the ground-state products. Hence, as the real reaction proceeds, the electronic energy of the system initially rises rapidly due to the $2A_{LHS} \rightarrow 3A_{RHS}$ correlation, but as the transition state is reached two electrons pass into the orbital described by the $3S_{LHS} \rightarrow 2S_{RHS}$ correlation. Examining the properties of the system at this point along the reaction coordinate (see $\times$ in Fig. 4-4), it is clear that the electronic energy of the system is high. Such concerted, symmetry-forbidden reactions should therefore occur with high barriers, as is observed experimentally.[5]

Four-center reactions (compare the above H_2 elimination) can occur without large reverse activation energies if they can occur in a relatively nonconcerted manner. Nevertheless, fragmentations which formally involve σ-bond formation appear to be frequently avoided due to large reverse activation energies, since their occurrence would otherwise be expected on thermodynamic grounds. For example, the saturated carbonium ion $C_5H_{11}^+$ undergoes C_2H_4 loss as the only significant metastable transition; this reaction is also the main reaction of $C_5H_{11}^+$ in the ion source. Relevant thermochemical data for possible reactants and product combinations are given in Fig. 4-5.[6]

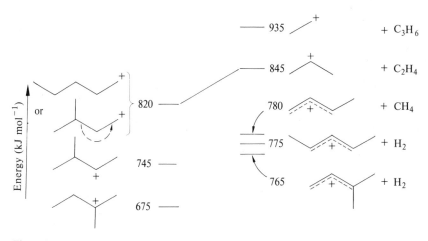

Figure 4-5

From Fig. 4-5, it is evident that (1) the $C_5H_{11}^+$ cation undergoes decomposition to give the most favorable combination of products which can arise by single-bond cleavage (see the broken arrow in Fig. 4-5); (2) the internal energies necessary to bring about this dissociation are sufficient to allow interconversion of primary, secondary, and tertiary $C_5H_{11}^+$ cations before dissociation; and (3) since H_2 and/or CH_4 losses from $C_5H_{11}^+$ remain hypothetical reactions in metastable transitions, these dissociations (involving σ-bond formation in the eliminated neutral) are concluded to involve reverse activation energies.

4-2 POTENTIAL ENERGY PROFILES

In Sec. 4-1A, it was seen that an ion may isomerize before fragmentation if the activation energy for isomerization is less than that for fragmentation of the original structure. Since it is then possible that fragmentation may occur from the isomeric structure, the fragmentation may not reflect the original structure.

The principles involved are illustrated by the behavior of the three isomeric ions, **10**, **11**, and **12**. Each of these is stable in the gas phase, existing in a potential well (Sec. 4-5). Each has the possibility to rearrange to another ion, given sufficient internal energy. The question is whether sufficient energy is available before fragmentation occurs.

$$CH_3CH{=}\overset{+}{O}H \qquad \overset{\displaystyle \overset{+}{O}H}{\underset{CH_2{-}CH_2}{\diagup \diagdown}} \qquad CH_2{=}\overset{+}{O}{-}CH_3$$

$$\textbf{10} \qquad\qquad\qquad \textbf{11} \qquad\qquad\qquad \textbf{12}$$

Metastable peaks (Sec. 3-2) give a direct picture of the kinetic energy released in a slow dissociation. It is found that **10** and **11** release essentially the same amount of kinetic energy when they dissociate by methane loss; the amount released is relatively large ($48 \, kJ \, mol^{-1}$) and relatively specific. Evidently, **10** and **11** dissociate over the same potential energy surface. Interconversion of **10** and **11** prior to their decomposition is also indicated by the observation that both ions, in metastable transitions, lose ca. 65% C_2H_2 and 35% CH_4. The fact that two ions dissociate through delicately balanced exit channels in very similar ratios is normally good evidence that they equilibrate before dissociation. This criterion, first applied in the study of $C_6H_{13}^{+[7]}$ and $C_2H_5O^{+[8]}$ ions is of sufficient general importance that it has become known as *the criterion of competing metastable transitions*.

In contrast, in metastable transitions **12** loses only CH_4, and with a much smaller (ca. $8 \, kJ \, mol^{-1}$) release of kinetic energy. Clearly, **12** dissociates without equilibration with, or conversion to, **10** or **11**.

Appearance energy measurements (Sec. 2-1A) establish that CH_4 losses from **10** and **12** require ca. 300 and $330 \, kJ \, mol^{-1}$, respectively. Knowing ΔH_f for the isomeric ions **10**, **11**, and **12**, these data allow a number of the levels on the potential energy profile, reproduced in Fig. 4-6, to be determined. The metastable peak shapes for the two processes for CH_4 loss are indicated schematically in the figure.

Using the data of Tables 4-1 and 4-4, ΔH_f for the hydroxyethyl cation can be estimated. This value cannot be easily determined directly since such a species is not anticipated to be in a potential well; ring closure to **11** by participation of the oxygen lone-pair or rearrangement to **10** via a 1,2-hydride shift is expected to occur without activation energy. Hence, a mechanistically plausible pathway for equilibration of **10** and **11** exists at internal energies less than those needed to cause dissociation. In contrast, perhaps the most facile route which can be envisaged for **10** or **11** → **12** involves C—C bond homolysis in **11** to give **13**. Assuming that the energy required for this process is given approximately by the C—C bond dissociation energy of cyclopropane, it is evident that **13** $(\Delta H_f \simeq 1040 \, \text{kJ mol}^{-1})$ is not accessible at the internal energies necessary to cause dissociation of **10, 11,** and **12.**

The potential energy profile so derived (Fig. 4-6)[9] is consistent with deuterium (^{2}H) and ^{13}C-labeling data for metastable transitions. These establish that (1) the oxygen-bound hydrogen of **10** remains attached to oxygen up to dissociation, (2) the carbon-bound hydrogens of **10** become equivalent, and (3) the carbon atoms of **10** become equivalent prior to unimolecular dissociations.

The available evidence indicates that potential energy profiles of the above form have general relevance to the unimolecular reactions which *any* ion may undergo in the mass spectrometer. Cases where one observed reaction pathway is not competitive with another observed pathway appear to be very much the exception rather than the rule. The observations that **14, 15, 16,** and **17** lose to some extent from their molecular ions CH_3, CO, C_2HD, and CH_3, respectively, establish in each case that there is an isomerization reaction of the initially generated molecular ion which requires less energy than its fragmentation; some isomeric structure then fragments.

The molecular ion from **16** fragments in a manner which suggests that the carbon atoms of the benzene ring have been reorganized (as can indeed be confirmed by ^{13}C-labeling). It would be quite wrong to conclude that the relative orientation of substituents in an aromatic ring cannot be determined by mass spectrometry. The isomerization of ionized **16** reflects its great stability toward

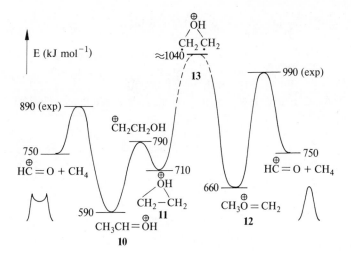

Figure 4-6

unimolecular fragmentation (requiring ca. 400 kJ mol^{-1}). In contrast, ionized **18** and **19** fragment much more readily and give characteristically different spectra (Sec. 6-4).

4-3 RATE-DETERMINING ISOMERIZATIONS

In many cases, it is found that isomerization of an ion A^+ to B^+ requires less energy than its direct dissociation, and that dissociation of B^+ is then more facile than reversion to A^+. That is, dissociation of A^+ always occurs via B^+ and the isomerization of A^+ to B^+ is the slow (rate-determining) step. The appropriate potential energy profile is given in Fig. 4-7.[10]

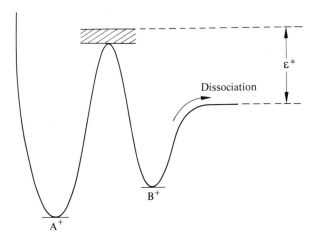

Figure 4-7

Experimental consequences of this situation are as follows:

1. The metastable peak for dissociation of B^+ formed from A^+ is broader than that for dissociation of directly generated B^+. This is caused by fluctuation of some of the excess energy $\varepsilon\ddagger$ into the reaction coordinate for dissociation of B^+. In other words, on a statistical basis, some of $\varepsilon\ddagger$ will emerge as relative translational energy of the dissociation products.
2. If directly generated, B^+ dissociates by two channels in competition to C^+ and D^+, then commencing from A^+ the $[C^+]/[D^+]$ metastable ratio will increase if the reaction $B^+ \rightarrow C^+$ has a higher frequency factor than $B^+ \rightarrow D^+$. This follows since higher internal energies favor reactions with higher frequency factors (Sec. 3-3).
3. Starting from A^+, a reaction sequence $A^+ \rightarrow B^+ \rightarrow E^+$ may be observed, although directly generated B^+ does not undergo the $B^+ \rightarrow E^+$ reaction. The measured appearance energy of C^+, D^+, and E^+, starting from A^+, will all be the same (within experimental error) and will in fact be a measure of the activation energy for the isomerization of $A^+ \rightarrow B^+$.

An example which illustrates the first of the above points is found in the rate-determining isomerizations of primary carbonium ions to more stable secondary or tertiary ions.[11] For example, the acylium ion **20** loses CO with release of a larger range of kinetic energies than does **21** (Fig. 4-8). The average kinetic energies released in these reactions (as computed from the metastable peak widths at half-height) are 0.4 (Fig. 4-8*a*) and 2.9 (Fig. 4-8*b*) kJ mol^{-1}.

Appearance energy measurements establish that unimolecular dissociation of **20** requires insufficient energy to produce the *n*-propyl cation together with CO. It is concluded that partial dissociation leads to **22**, in which the incipient carbonium ion then rearranges to the complex **23** containing the more stable secondary cation. This is in accord with other experimental and theoretical findings that primary carbonium ions isomerize to more stable secondary forms essentially without activation energy. The potential energy which is released in the process **22 → 23** occasionally fluctuates into the reaction coordinate for CO loss, as seen in Fig. 4-8.

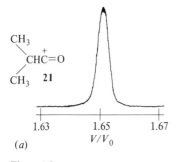

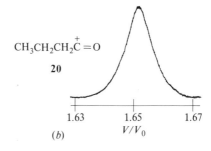

Figure 4-8

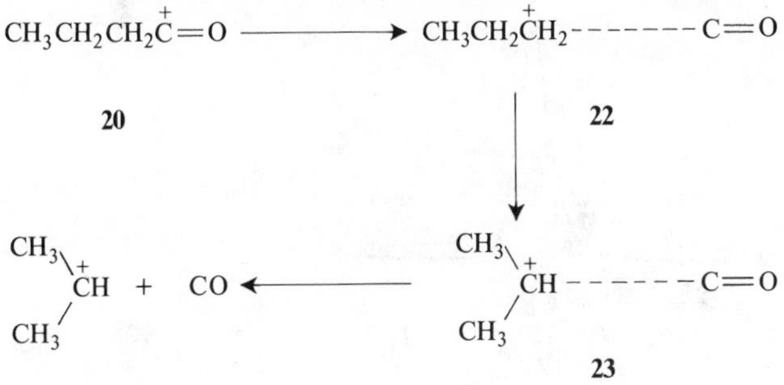

$$CH_3CH_2CH_2\overset{+}{C}{=}O \longrightarrow CH_3CH_2\overset{+}{C}H_2 \text{-------} C{=}O$$

20   **22**

$$\underset{CH_3}{\overset{CH_3}{>}}\!\!\overset{+}{C}H \;+\; CO \longleftarrow \underset{CH_3}{\overset{CH_3}{>}}\!\!\overset{+}{C}H \text{-------} C{=}O$$

23

Carbon monoxide has a very small dipole moment (0.1 Debye) and therefore ion dipole attractive forces in the complexes **22** and **23** are relatively weak. Semiquantitatively, the ion-dipole attraction E is given by

$$E = \frac{q\mu\cos\theta}{r^2} \tag{4-6}$$

where q is the charge on the ion, μ is the dipole moment of the coordinated molecule, r is the distance between the point charge and point dipole, and θ is the angle between the dipole and the charge to dipole axis.

If CO as leaving group is replaced by formaldehyde, which has a much larger dipole moment (2.3 Debye), then the analogous dissociations are depicted by **24** to **27**.

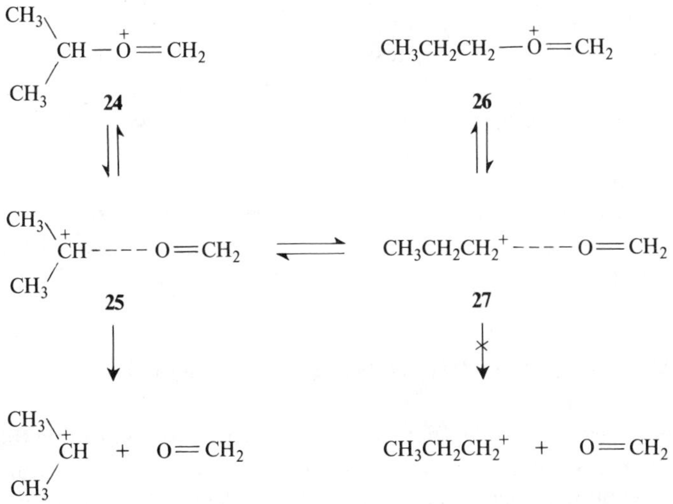

$$\underset{CH_3}{\overset{CH_3}{>}}\!\!CH-\overset{+}{O}{=}CH_2 \qquad\qquad CH_3CH_2CH_2-\overset{+}{O}{=}CH_2$$

24   **26**

$$\underset{CH_3}{\overset{CH_3}{>}}\!\!\overset{+}{C}H\text{----}O{=}CH_2 \;\rightleftharpoons\; CH_3CH_2\overset{+}{C}H_2\text{----}O{=}CH_2$$

25   **27**

$$\underset{CH_3}{\overset{CH_3}{>}}\!\!\overset{+}{C}H \;+\; O{=}CH_2 \qquad CH_3CH_2\overset{+}{C}H_2 \;+\; O{=}CH_2$$

If, at a $RCH_2{}^+$---$O{=}CH_2$ bond length of 0.3 nm, the incipient n-propyl cation is sufficiently free to rearrange to the complex **25**, it will rearrange when the ion-dipole binding energy of **27** is $\sim 75\,kJ\,mol^{-1}$. The potential energy released by **27** $\rightarrow$ **25** will be equal to or (more probably) slightly less than that ($65\,kJ\,mol^{-1}$) for isomerization of a primary to a secondary carbonium ion. Thus, the analogous

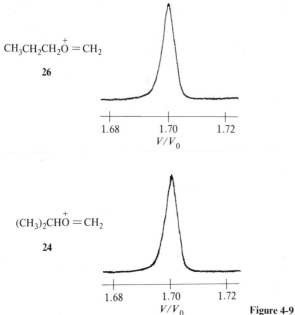

$$CH_3CH_2CH_2\overset{+}{O}=CH_2$$

26

1.68 1.70 1.72
V/V_0

$$(CH_3)_2\overset{+}{CHO}=CH_2$$

24

1.68 1.70 1.72
V/V_0 **Figure 4-9**

ion-dipole attraction in **25** is likely to be sufficient to prevent dissociation of **25**, despite its initial production with $\sim 65\,kJ\,mol^{-1}$ of excess internal energy. Thus, **25** and **27** may equilibrate more readily than they dissociate. This is indeed the case, as indicated by the fact that the dissociation of both **24** and **26** occurs with the same release of kinetic energy (Fig. 4-9). The metastable peaks reproduced in Fig. 4-9 (and Fig. 4-8) are due to dissociations occurring in the first field-free region of a double-focusing mass spectrometer and were obtained by scanning the accelerating voltage (Sec. 3-2B).

In analogy to solution chemistry, it may be said that the strong "solvation" of the carbonium ions by formaldehyde makes equilibration of the cations fast relative to dissociation.

4-4 PRIMARY DEUTERIUM ISOTOPE EFFECTS

It has been shown (Sec. 3-3C) that reactions occurring as metastable transitions do so with little excess energy in the transition state. Hence, such reactions might be expected to exhibit primary deuterium isotope effects when the rate-determining step involves the stretching of an X—D bond in competition with stretching of a weaker X—H bond. In solution reactions, such isotope effects (measured in terms of k_H/k_D) often lie in the range 1.5–3. Some analogous primary deuterium isotope effects observed in metastable transitions (and measured from the relative abundance of these transitions) are given in Table 4-6.

Table 4-6 Some primary deuterium isotope effects observed in metastable transitions

Number	Precursor ion	Competing losses	Isotope effect†
1	CHD_3^+	$H \cdot / D \cdot$	$H \cdot$ loss only
2	$HDC{=}CDH^+$	H_2/D_2	$\sim 700{:}1$
3	$CH_3CD_3^+$	$H \cdot / D \cdot$	$600{:}1$
4	$C_2H_2D_3^+$	H_2/D_2	$8{:}1$
5	$C_3H_4D_3^+$	H_2/D_2	$2.6{:}1$
6	$C_6H_2D_5^+$	H_2/D_2	$3{:}1$
7	$C_6H_5CD_3^+$	$H \cdot / D \cdot$	$2.8{:}1$

† These values have been multiplied by an appropriate statistical factor where relevant, in order to correspond to competition where equal numbers of H and D atoms are available.

In Table 4-6, the precursor ions for (1), (2), (3), and (7) are available by ionization of partially deuteriated methane, ethylene, ethane, and toluene, respectively. The partially deuteriated ethyl (4) and propyl cations (5), and "protonated benzene" cations (6), are available by fragmentation of suitable precursors. The ions (2), (4), (5), (6), and (7) are all established to undergo, prior to unimolecular decomposition, rearrangements which allow both H and D atoms to become statistically distributed in the ions.

It is clear that in metastable transitions (i) for small ions the primary deuterium isotope effects are large,[12] much larger than those found in solution reactions, and (ii) for medium-sized organic ions the isotope effects are comparable to those found in solution chemistry[13, 14] (see also Sec. 11-3).

Theory predicts that large isotope effects should be observed at ion energies just above the threshold for decomposition (see also Sec. 11-3). Very large isotope effects are expected for the metastable transitions from small ions, because the energy range giving metastables for such ions is often extremely narrow due to the rapid rise of k with E (compare Fig. 3-9). If partial deuteriation introduces a new reaction, e.g., loss of D whose activation energy is ca. $8\,kJ\,mol^{-1}$ higher than that for loss of H, then the high-energy reaction might almost fall outside the metastable window. The data in Table 4-6 [entries (1) to (3)] bear out this point. For larger ions, the necessary activation energy is dispersed in more vibrational modes and fluctuates into the reaction coordinates with much lower probability. Hence, k rises much more slowly with E, and the isotope effects are smaller (Table 4-6).

It may seem surprising that some of the isotope effects given in Table 4-6 are so much bigger than those observed in solution reactions. However, it must be remembered that the competition is determined by the excess energy in the transition state; in a solution, reactions associated with high-energy collisions must be integrated with those associated with less excess energy.

4-5 COLLISIONAL ACTIVATION

It is possible to raise the pressure in a field-free region of a mass spectrometer by the introduction of a gas, e.g., hydrogen, helium, or air. Ions in the ion beam with large translational energies (several thousand electronvolts acquired during ion acceleration) collide inelastically with the neutral atoms or molecules. As a consequence, part of the translational energy of the ion may be converted into internal energy, which subsequently causes decomposition of the ion.[15-17] The pattern of decomposition undergone by an ion upon collision is termed its *collisional activation* (CA) spectrum.

The CA spectrum of an ion can provide a fingerprint which is characteristic of its structure. The CA spectrum may be conveniently obtained by introducing the collision gas into the second field-free region of a reverse-geometry instrument (Sec. 1-3A). The ion of interest is selected by an appropriate setting of the magnetic field of the first sector; all fragments produced upon collision in the field-free region following the magnetic sector are recorded by scanning the voltage E in the second (electric) sector. The spectrum obtained in this way is in fact a superposition of collision-induced and unimolecular metastable ion dissociations. The CA spectrum is frequently similar to the normal 70 eV electron-impact spectrum. The time available for fragmentation in a field-free region is similar to that available for fragmentation in a conventional electron-impact source. Hence, we may conclude that the energy distributions in the two techniques are similar.

In the present context, it is important to realize that the CA fingerprint of an ion is obtained mainly from those ions which have been given insufficient energy in the mass spectrometer source to dissociate. Consider four isomeric ions A^+, B^+, C^+, and D^+ (Fig. 4-10) where the potential energy profile is such that only A^+, C^+, and D^+ undergo direct unimolecular dissociation, as indicated; all other barriers are to isomerization.

The unimolecular metastable ion (MI) spectra of A^+, B^+, and C^+ will be essentially the same. These spectra, conveniently obtained by a MIKES or B/E scan (Sec. 1-3A), will show competing dissociations (via structures A^+ and C^+).

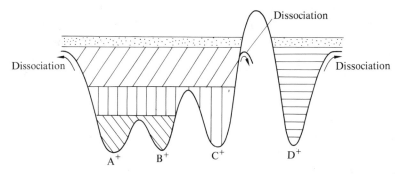

Figure 4-10

The MI spectrum of D^+ will be distinct. The ions in all the MI spectra arise from precursor ions with internal energies indicated by ⊛.

The ions D^+ which are energized by collision have, before collision, energies indicated by ⊖. These energies are raised to a large range of higher energies by collision, some of which will allow isomerization to C^+, B^+, and A^+. However, some collision-induced dissociations of D^+ will be so fast that prior rearrangement of D^+ is precluded. Thus, a considerable portion of the CA spectrum of D^+ will be characteristic of the D^+ structure.

The CA spectra of ions A^+, B^+, and C^+ will be derived from all ions in the areas indicated by ⊘, ⦸, and ⦶. That of C^+ is therefore expected to be distinct from those of A^+ and B^+. However, the CA spectra of A^+ and B^+ may be barely (and not securely) distinguishable since, of the nondecomposing A^+ and B^+ ions selected for analysis, only those in the ⊘ area are precluded from interconverting. In general, if two ions are initially generated with isomeric structures but with no energy barrier to conversion of one to the other, then these ions will produce the same CA spectra.

A further point, which can be illustrated by reference to Fig. 4-10, is the relationship of ion structure to internal energy. Although energetically unexcited A^+, B^+, C^+, and D^+ have uniquely defined structures, A^+ which dissociates in metastable transitions can rearrange rapidly and reversibly to B^+ and C^+ during its flight to the field-free region. At times it is also appropriately defined as a transition state containing excess energy between B^+ and C^+. The population of this state is, however, much less than that of A^+, B^+, and C^+ since for a given total internal energy those structures with the maximum nonfixed energy (minimum potential energy) are most highly populated (Sec. 2-1C).

The foregoing discussion demonstrates that distinct CA spectra can be used to indicate which of a series of ions, believed to be initially generated as isomeric structures, exist in discrete potential energy wells. Data for isomeric $C_2H_5O^+$ ions are given in Table 4-7.

It has been surmised earlier (Fig. 4-6) that each of the ions **10**, **11** and **12** exists

Table 4-7 Partial collision activation spectra of $C_2H_5O^+$ ions[18]

Precursor	Ion structure		Relative abundance, m/z					
			24	25	26	28	30	31
$CH_3OCH_2CH_2OCH_3$	$CH_3O^+{=}CH_2$	(**12**)	<1	<1	4	41	43	12
$CH_3OCH_2CH_2CN$	$CH_3O^+{=}CH_2$	(**12**)	<1	2	4	39	43	12
CH_3CHO†	$CH_3CH{=}\overset{+}{O}H$	(**10**)	4	16	45	18	14	3
$(CH_3)_2CHOH$	$CH_3CH{=}\overset{+}{O}H$	(**10**)	4	14	49	15	14	4
$CH_3CH_2(CH_3)CHOH$	$CH_3CH{=}\overset{+}{O}H$	(**10**)	5	16	47	17	12	3
$\underline{CH_2CH_2O}$†	$\underline{CH_2CH_2\overset{+}{O}H}$	(**11**)	4	15	42	13	11	16

† By protonation in a source operating at higher pressure.

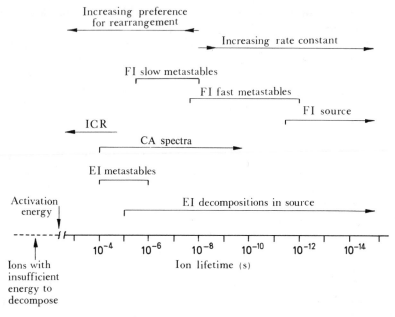

Figure 4-11

in a potential energy well. This conclusion is supported by the three distinct sets of CA spectra (Table 4-7).

The discussion of CA spectra completes the consideration of a number of techniques for examining unimolecular reactions. The various lifetimes appropriate to these techniques are illustrated as a general scheme in Fig. 4-11.

4-6 THE STEVENSON–AUDIER RULE

It is relevant to the energetics of ion formation, and also a suitable prelude to the relationship between fragmentation and structure (Chap. 6), to discuss a useful rule concerned with the prediction of preferred fragmentations. In connection with the fragmentation of alkanes in the mass spectrometer, Stevenson noted that the positive charge remained on the more substituted fragment,[19] i.e., the one with the lower ionization energy. This principle may be extended to all mass-spectral fragmentations, and it may be generally stated that the positive charge will remain on the fragment of a lower ionization energy.[20] Since the idea arose from the work of two authors,[19, 20] we have called it the Stevenson–Audier rule.

This principle may be simply explained by reference to the energy diagram shown in Fig. 4-12, which illustrates the energy changes involved in production of A^+ and B^+ from a molecule AB (where A and B are any radicals). The symbols IE and AE have the usual meaning and $D(A—B)$ refers to the dissociation energy of

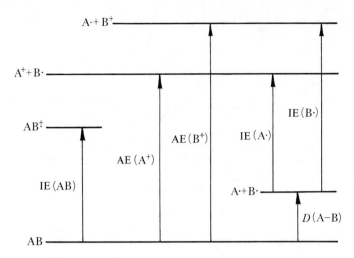

Figure 4-12

the bond connecting the two radicals. Any excess energies or reverse activation energies are neglected.

It is evident from Fig. 4-12 that the lower AE reaction (formation of A^+ in this case) corresponds to the process where the charge resides on the fragment of lower IE [IE(A $\cdot$) < IE(B $\cdot$)]. The rule does not predict which of all the possible fragment ions will be the most prevalent, but indicates which way the charge will go for fragmentation at a given bond.

Consider the ionization energies of some hydrocarbon radicals reproduced in Table 4-8.[21] Two trends are evident from these figures. First, there is a tendency for the IE to decrease as molecular size increases for radicals of the same type (e.g., primary radicals). Thus, for example, formation of $C_8H_{17}^+$ would be predicted to be a lower-energy process than formation of $C_2H_5^+$ from the molecular ion of n-decane ($C_{10}H_{22}^+$) (especially if hydrogen rearrangement can occur in formation of $C_8H_{17}^+$; see Secs. 4-1 and 4-3). This is borne out in practice, and a metastable peak is observed for the former but not for the latter process.

The second important tendency is the lowering of the ionization energy as the radical becomes more highly substituted. The reason for the preferential formation of secondary or tertiary hydrocarbon ions in the spectra of branched hydrocarbons is therefore evident.

Figure 4-12 may be modified to account for rearrangement reactions of molecular ions with elimination of stable neutrals (see Fig. 4-13). The sum $\Sigma D(M)$ represents the difference between the sums of the heats of bond rupture and formation in the rearrangement reaction $M \rightarrow C + D$. The treatment of rearrangement reactions as in Fig. 4-13 makes simplifying assumptions in ignoring reverse activation energies, but works well in practice.

The retro-Diels–Alder reaction provides a suitable illustration of the application of the Stevenson–Audier rule. For example, the molecular ion of α-terpineol

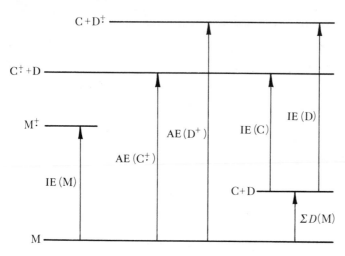

Figure 4-13

Table 4-8 Ionization energies of hydrocarbon radicals

Radical	IE, eV	Radical	IE, eV
$CH_3 \cdot$	9.84	$CH_3CH_2CH_2CH_2 \cdot$	8.01
$CH_3CH_2 \cdot$	8.38	$(CH_3)_2CHCH_2 \cdot$	8.01
$CH_3CH_2CH_2 \cdot$	8.10	$CH_3CH_2CH \cdot CH_3$	7.41
$(CH_3)_2CH \cdot$	7.55	$(CH_3)_3C \cdot$	6.93

(**28**) can decompose via a retro-Diels–Alder reaction to form the diene **29** and the allylic alcohol **30**. The relative ionization energies of the two even-electron species (**29**, 9.0 eV; **30**, 9.4 eV) lead to the prediction that the peak due to the charged diene will be the more prevalent in the spectrum, and this is observed to be the case.

Trans-Δ^2-octalin (**31**), however, yields a mass spectrum having peaks at m/z 54 and m/z 82 of almost equal abundance. Cyclohexene (**32**) and butadiene (**33**) both possess ionization energies close to 9.1 eV.

Examples are also forthcoming from the field of substituent effects. A γ-hydrogen (McLafferty) rearrangement from the γ-substituted methyl butyrate (**34**) may yield either the $M^+ - 74$ ion (**35**) or the m/z 74 ion (**36**).

When $R = NO_2$, ion **36** is the base peak of the spectrum, whereas the abundance of ion **35** is low. When $R = OMe$, the situation is reversed. When $R = H$, **35** and **36** have similar abundances. This effect is merely a consequence of the lowering of the ionization energy of the neutral compound **35** as the substituent R becomes more electron donating.

REFERENCES

1. H. M. Rosenstock, K. Draxl, B. W. Steiner, and J. T. Herron: *J. Phys. Chem. Ref. Data*, vol. 6, suppl. 1, 1977; for a detailed discussion of the energetics of unimolecular reactions, see K. Levsen: "Fundamental Aspects of Organic Mass Spectrometry," Verlag Chemie, Weinheim, 1978.
2. J. L. Franklin: *J. Chem. Phys.*, vol. 21, p. 2029, 1953.
3. L. Radom, J. A. Pople, and P. von R. Schleyer: *J. Am. Chem. Soc.*, vol. 94, p. 5935, 1972.
4. R. B. Woodward and R. Hoffmann: *Agnew. Chem. Int. Edn. Eng.*, vol. 8, p. 797, 1969.
5. D. H. Williams and G. Hvistendahl: *J. Am. Chem. Soc.*, vol. 96, p. 6753, 1974.
6. R. D. Bowen and D. H. Williams: *J. Chem. Soc. Perkin II*, p. 1479, 1976.
7. H. M. Rosenstock, V. H. Dibeler, and F. N. Harllee: *J. Chem. Phys.*, vol. 40, p. 591, 1964.
8. T. W. Shannon and F. W. McLafferty: *J. Am. Chem. Soc.*, vol. 88, p. 5021, 1966.
9. R. D. Bowen, D. H. Williams, and G. Hvistendahl: *J. Am. Chem. Soc.*, vol. 99, p. 7509, 1977.
10. D. H. Williams: *Accts. Chem. Res.*, vol. 10, p. 280, 1977.
11. D. H. Williams, B. J. Stapleton, and R. D. Bowen: *Tetrahedron Letters*, p. 2919, 1978.
12. U. Löhle and Ch. Ottinger: *J. Chem. Phys.*, vol. 51, p. 3097, 1969.
13. I. Howe and F. W. McLafferty: *J. Am. Chem. Soc.*, vol. 93, p. 99, 1971.
14. H. Eyring and F. W. Cagle: *J. Phys. Chem.*, vol. 56, p. 889, 1952.
15. K. R. Jennings: *Int. J. Mass Spec. Ion Phys.*, vol. 1, p. 227, 1968.
16. F. W. McLafferty, P. F. Bente, R. Kornfeld, S-C. Tsai, and I. Howe: *J. Am. Chem. Soc.*, vol. 95, p. 2120, 1973.
17. K. Levsen and H. Schwarz: *Angew. Chem. Int. Edn. Eng.*, vol. 15, p. 509, 1976.
18. B. van de Graf, P. P. Dymerski, and F. W. McLafferty: *Chem. Comm.*, p. 978, 1975.
19. D. P. Stevenson, *Disc. Faraday Soc.*, vol. 10, p. 35, 1951.
20. See, for example, H. E. Audier: *Org. Mass Spec.*, vol. 2, p. 283, 1969.
21. F. P. Lossing and G. P. Semeluk: *Can. J. Chem.*, vol. 48, p. 955, 1970.

BIMOLECULAR REACTIONS

In the previous chapter, the principles governing the unimolecular reactions of organic ions are discussed; these are reactions of isolated cations without any interference from other molecules. This situation differs fundamentally from that pertaining in solution because all effects arising from interactions between reactant ions and solvent molecules are preempted. An intermediate situation is encountered in the bimolecular reactions of ions. After the ions are generated, usually in one of the conventional ways (Sec. 1-2), they are allowed to come into contact with neutral molecules. An ion-molecule reaction may then occur, often leading to the formation of a new ion and neutral. The product ion, or ions, are subsequently characterized by mass analysis, frequently using a quadrupole mass filter (Sec. 1-3B).

As indicated in Chap. 1, the pressure (ca. 10^{-7} torr) in conventional mass spectrometers is too low to permit collisions to take place to a significant extent. In order to observe bimolecular processes, it is necessary to allow collisions to occur; this is normally achieved in one of two ways. Either the low pressure is maintained, but longer ion lifetimes are employed, or higher pressures are used. In the former case, the ions survive for relatively long periods (milliseconds or longer); consequently, there is enough time for interactions to take place between ions and neutral molecules admitted into the system. In the case of higher pressures (typically in the range 0.1 to even 100 torr), bimolecular reactions may be observed without increasing the ion lifetime.

Equilibrium may be established between the forward and reverse reactions for a given bimolecular process:

$$A^+ + B \rightleftharpoons C^+ + D \tag{5-1}$$

Furthermore, on account of the collisions between molecules, a Maxwell–Boltzmann distribution is set up for the internal energies of reactants and products.

As a result, thermodynamic parameters may be measured; in particular, the temperature T of the system may be determined. This is in stark contrast to unimolecular reactions, where the concept of temperature is essentially meaningless.

The rates of both forward and reverse reactions may be obtained; in addition, it is also possible to measure the equilibrium constant (K) for the overall process. These experiments may be performed for a range of different temperatures; the enthalpy change (ΔH) associated with the reaction may then be calculated from a Van't Hoff plot of $\log K$ against $1/T$.

5-1 MECHANISMS OF ION-MOLECULE REACTIONS

It is pertinent to give a brief account of some of the current views on the mechanisms of ion-molecule reactions. This summary is designed to present only the general principles governing ion-molecule reactions involving organic ions. More sophisticated mathematical treatments have been developed to explain in detail the mechanisms and rates of some reactions involving very small molecules and ions;[11] however, such analyses are beyond the scope of this chapter.

Bimolecular rate constants for a number of processes are given in Table 5-1. From these representative data, it is evident that ion-molecule reactions in the gas phase have higher rates than reactions between neutral species. This is a general trend and arises from several causes, two of which are discussed below.

Table 5-1 Examples of bimolecular rate constants

Reaction	k ($cm^{-1}\,mol^{-1}\,s^{-1}$)	T (°C)
Gas-phase ion-molecule reactions		
$Ar^{+} + H_2 \rightarrow ArH^{+} + H\cdot$	3.5×10^{-10}	25
$H_2O^{+} + H_2O \rightarrow H_3O^{+} + HO\cdot$	1.3×10^{-9}	25
$CH_4^{+} + CH_4 \rightarrow CH_5^{+} + CH_3\cdot$	1.2×10^{-9}	100
$CH_2NH_2^{+} + CH_3NH_2 \rightarrow CH_3NH_3^{+} + CH_2NH$	2.4×10^{-9}	100
$C_2H_2^{+} + C_2H_3F \rightarrow C_2H_3F^{+} + C_2H_2$	1.9×10^{-9}	100
$CH_3OCD_3^{+} + CH_3OCD_3 \rightarrow CH_3OCD_3H^{+} + \cdot CH_2OCD_3$	1.15×10^{-9}	100
$C_4H_2^{+} + C_6H_6 \rightarrow C_{10}H_7^{+} + H\cdot$	4.7×10^{-10}	$\cdots$
Gas-phase reactions involving radicals		
$C_2H_5\cdot + C_2H_5\cdot \rightarrow C_4H_{10}$	2.7×10^{-11}	25
$CH_3\cdot + C_2H_6 \rightarrow CH_4 + C_2H_5\cdot$	8×10^{-22}	25
$Br\cdot + H_2 \rightarrow HBr + H\cdot$	8×10^{-24}	25
Gas-phase reactions between molecules		
$NO + O_3 \rightarrow NO_2 + O_2$	2×10^{-14}	25
$HI + C_2H_5I \rightarrow C_2H_6 + I_2$	1.6×10^{-28}	25
$NO_2 + NO_2 \rightarrow 2NO + O_2$	1.6×10^{-31}	25
Ions in aqueous solution		
$CO_2 + OH^{-} \rightarrow HCO_3^{-}$	7×10^{-18}	25
$CH_3COOC_2H_5 + OH^{-} \rightarrow CH_3COO^{-} + C_2H_5OH$	1.6×10^{-22}	25

First, most ion-molecule reactions occur with relatively small activation energies. This reflects the fact that the ion and neutral often can associate exothermically to form a collision complex. In this connection, it is instructive to relate the rate (k_1) of ion-molecule reactions to the collision frequency. The simple theory of molecule collisions, as applied to bimolecular reactions, predicts the following rate constant:

$$k_1 = 2\sigma^2 \left(\frac{2\pi kT}{\mu}\right)^{1/2} e^{-E_a/kT} \tag{5-2}$$

where σ is the collision diameter, μ is the reduced mass of the reactants [$\mu = m_1 m_2/(m_1 + m_2)$], k is Boltzmann's constant, and E_a is the activation energy.

If the reaction has no activation energy, the expression for k_1 reduces to the preexponential term in Eq. (5-2). This residual expression is a maximum value, because the derivation of Eq. (5-2) assumes that every collision results in reaction. In fact, many ion-molecule reactions require a specific orientation of the reactants; consequently, some, or many, collisions may lead to no reaction. Therefore, the experimental value for k_1 could in principle be considerably less than the maximum predicted on the basis of Eq. (5-2). Reasonable values for the parameters in Eq. (5-2) are: $\sigma = 10^{-7}$ cm, $T = 373$ K, $m_1 = m_2 = 40/(6 \times 10^{23})$ g; substituting in Eq. (5-2) yields a value for k_1 of 2×10^{-9} cm^3 mol^{-1} s^{-1}.

Thus, from Table 5-1, it appears that many ion-molecule reactions in the gas phase exhibit the maximum rate constant, i.e., most collisions result in reaction, within the limitations of the hard-sphere model used to derive Eq. (5-2).

The rapidity of ion-molecule reactions may be understood in terms of the Langevin model.[2] Long-range attractive forces are produced between the approaching ion and molecule due to induction of polarization in the target molecule (or increased polarization, if a permanent electric dipole is already present). Depending on the proximity and relative velocity of the two species, these attractive forces can cause the distance of closest approach to be sufficiently small for reaction to occur. Moreover, when the ion approaches the target molecule within a certain range of distances, the trajectory of the ion takes on a spiral, or orbiting, nature around the molecule. In such cases, the duration of the interaction is greatly increased (see Fig. 5-1a). However, on account of short-range repulsive

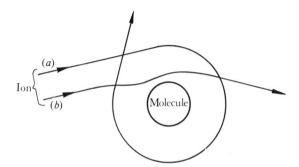

Figure 5-1

forces, a lower limit exists for the distance of closest approach, which leads to these extended interactions. Ions which approach on a still closer trajectory do not enter the metastable orbit (see Fig. 5-1b).

In summary, the simplified approach outlined above shows that the cross section for an ion-molecule reaction is increased. This is because of (1) the long-range attractive forces between the two species and (2) the possible orbiting interaction between the ion and molecule, producing a collision complex which prolongs the duration of the interaction.

Evidence is available which indicates that the collision complex may have a lifetime which is long compared with vibrational and rotational periods, thus permitting extensive reorganization to take place prior to reaction. An example is furnished by the elimination of H_2 from the collision complex $[C_2H_7]^+$, produced by association of CH_3^+ and CH_4; almost complete loss of identity of all seven hydrogen atoms occurs before H_2 is lost, as demonstrated by deuterium-labeling experiments.[3]

5-2 METHODS FOR OBSERVING BIMOLECULAR REACTIONS

Three essential conditions must be satisfied in order to observe bimolecular ion-molecule processes under equilibrium conditions.

1. The entire reaction system must be in equilibrium with the surrounding apparatus. This means that the ions and neutral molecules must be in thermal equilibrium with the walls of the reaction vessel and any carrier gas molecules which may also be present. In practice, the neutral molecules are present in very high concentration compared to the ions. Moreover, these neutral molecules are usually stable and may be introduced into the system with energies comparable to that appropriate to equilibrium. Consequently, the neutral molecules normally become thermalized very rapidly. In contrast, the ions may well be produced with excess energy; in addition, further energy may be imparted to the ions, after formation, by the application of accelerating or collimating electric potentials. Therefore, it is necessary to ensure that there are sufficient collisions between the ions and other molecules to allow the ions to become thermalized.
2. The rate at which both reactant and product ions undergo the reaction of interest must be fast, compared to the rates at which these ions can react via other processes. This condition reflects the requirement that the concentrations of the reactant and product ions must depend essentially only on the nature of the equilibrium under investigation.
3. The system must be given enough time to attain equilibrium.

Three general types of instrumentation are employed in practice to study equilibrium bimolecular processes; these are the ion-cyclotron resonance spectrometer, the flowing afterglow apparatus, and the high-pressure ion source.

A. Ion-Cyclotron Resonance

The principles of ion-cyclotron resonance (ICR) are discussed in Sec. 1-3D. Even at very low pressures (for example, 10^{-7}–10^{-4} torr), ion-molecule reactions are detected because of the greatly increased ion lifetimes (typically milliseconds, though much longer lifetimes are feasible).[4] Using the double-resonance technique,[5] it is possible to discover which daughter ions are produced by ion-molecule reactions of a given parent ion (Sec. 1-3D). Furthermore, proton-transfer processes involving either anions or cations may be studied by means of the pulsed double-resonance procedure.[6] Thus, a scale of gas-phase acidities or basicities may be established. Moreover, if a suitable thermochemical reference point can be found with which to calibrate this scale, absolute values may be assigned for the proton and hydride affinities of the relevant species. These thermochemical data may then be employed to deduce accurate heats of formation for isolated ions.[7]

It is often found that the order of acidity (or basicity) encountered in the gas phase is different from that found in solution. Examples of this behavior are presented in Sec. 5-3.

B. Flowing Afterglow

As pointed out in the introduction, ion-molecule reactions may be observed by increasing either the ion lifetime or the pressure in the reaction apparatus. The former approach is used in ICR whereas the latter is utilized in the flowing afterglow apparatus and the high-pressure ion source.

A schematic diagram of a flowing afterglow apparatus is depicted in Fig. 5-2.[8] Ions are produced by conventional means (usually electron-impact ionization is employed) and swept through the flow tube by means of a carrier gas (hydrogen and helium are often used). The flow rate in the tube is adjusted so that ions undergo many collisions with the molecules of the carrier gas, and thus become thermalized, prior to reacting with the neutral molecules of interest. Suitable reactant molecules can be introduced into the system through gas inlet ports, which may be either fixed or movable. Subsequently, some ions enter the quadrupole mass filter, which is situated at the end of the flow tube, and are subjected to mass analysis. Ion-molecule reactions of selected neutrals with a given

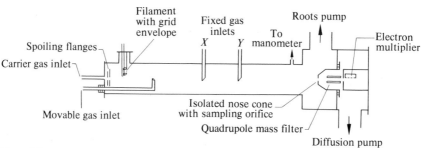

Figure 5-2

reactant ion may then be investigated. The temperature of the reaction medium may be varied over a wide range (100–600 K) by enclosing the entire apparatus in a suitable cooling or heating mantle.[9] In addition, the pressure may be controlled by altering the flow rate of the carrier gas; this pressure is normally monitored, using a precision capacitance manometer pressure gauge located near the middle of the reaction region. Typical pressures lie in the range 0.1–3 torr, with values between 0.5 and 1 torr being most usual.

The rate of the relevant bimolecular reaction is conveniently measured by determining the amount of reagent gas needed to reduce the parent ion signal by one order of magnitude. However, sophisticated analyses are often required to extract the best value for the bimolecular rate constant.

Consider a given bimolecular process [Eq. (5-1)] with an associated equilibrium constant K. Suppose $[A^+]$ and $[C^+]$ are the concentrations of the reactant and product ions, respectively, and let the flow rates of the respective neutrals be F_B and F_D; in principle, K is then given by

$$K = \frac{[C^+]F_D}{[A^+]F_B} \tag{5-3}$$

The flow rates F_B and F_D are known; $[A^+]$ and $[C^+]$ may be determined from the mass analysis results, after application of a mass discrimination factor. Consequently, K may be obtained from the slope of the plot of $[C^+]F_D$ against $[A^+]F_B$.

The temperature variation of K may be utilized to measure the heat of reaction (ΔH) for the overall process. This permits the determination of proton affinities and other thermochemical data, which may be compared with results obtained from ICR work.

Gas-phase substitution reactions may be investigated using the flowing afterglow approach. Fundamental differences are found between the situations pertaining in solution and the gas phase; these results are discussed in Sec. 5-4.

C. High-Pressure Ion Sources

It is possible to design ion sources which operate at relatively high pressures (for example, 1–10 torr). These sources are not operated such that ionization occurs continuously; instead, a short period in which ionization takes place is followed by a longer period in which the ionization process is discontinued. Consequently, these sources are referred to as pulsed high-pressure sources.

In a typical example of this variety of source,[10] ionization is effected using a pulse of high-energy (2000 eV) electrons. Each pulse lasts for some 10 μs and contains approximately 10^6 electrons; after the pulse, the electron beam is switched off for 6 ms and the pulse cycle is then repeated. The precursor molecules, used to generate the ion in question, are allowed to flow through the ion source in the presence of a carrier gas (methane is often used). Ions produced during the pulse of electrons then undergo collisions with the neutral molecules of interest. Some ions subsequently escape from the source through a leak and are subjected to mass analysis.

The system is especially suitable when the reactant ion and neutral molecule can both be produced from the same reagent gas. For example, H_3O^+ may be formed by bombarding H_2O in a stream of methane carrier gas; ion-molecule reactions of H_3O^+ with H_2O can then be investigated.[10]

5-3 ACID-BASE BEHAVIOR

Acid-base equilibria may be studied using any of the three experimental methods outlined in the previous section. Processes involving positive or negative ions may be investigated:

$$WH^+ + X \rightleftharpoons W + XH^+ \tag{5-4}$$

$$Y^- + ZH \rightleftharpoons YH + Z^- \tag{5-5}$$

The order of acidity (or basicity) is frequently different in the gas phase from that observed in solution. This phenomenon indicates that the effects of solvation in governing the relative acidity of species in solution are often highly significant. Solvation effects cannot be ignored or dismissed as being of secondary importance. Three examples are given below to illustrate this point.

The relative acidities of numerous aliphatic alcohols can be determined by examining the equilibrium

$$ROH + R'O^- \rightleftharpoons RO^- + R'OH \tag{5-6}$$

These experiments[6] show that the order of acidity is: neopentyl alcohol > t-butyl > isopropyl > ethyl > methyl > water; also, t-butyl $\approx$ n-pentyl $\approx$ n-butyl > n-propyl > ethyl. The order obtained for gas-phase acidities is the reverse of that found for the acidities of the alcohols in solution. Thus, in contrast to solution behavior, t-butoxide anion is a weaker base than methoxide anion in the gas phase.

These results undermine the traditional rationalization for the high basicity of t-butoxide anion in solution; namely, that the electron-rich methyl groups increase the charge density on oxygen by inductive (or hyperconjugative) electron donation. This explanation invokes trends in the intrinsic properties of alkoxide anions; consequently, the same trend would be evident in the gas phase. Hence, the high basicity of t-butoxide is due to solvent effects. The order of basicity in the gas phase may be interpreted in terms of the differences in polarizability of the alkyl groups. The larger t-butyl group can stabilize the negative charge in the corresponding alkoxide anion more effectively by induced polarization than the smaller and less-polarizable methyl group. In solution, this trend is outweighed by hydrogen bonding, which occurs more effectively for smaller ions, thus reversing the order of basicity of the alkoxide ions. Solvent molecules can readily envelop and stabilize the CH_3O^- ion by hydrogen bonding; however, in $(CH_3)_3CO^-$, the bulky t-butyl group partially shields the negative charge from interaction with solvent molecules.

A second example of the importance of solvent effects in influencing the relative basicity of homologous molecules is furnished by the ease of protonation of primary, secondary, and tertiary amines. In the methyl-substituted series, the order of basicity found in solution is $(CH_3)_2NH \gtrsim CH_3NH_2 > (CH_3)_3N > NH_3$. Attempts have been made to rationalize this apparently anomalous order in terms of electronic and steric effects. On a naive basis, electron donation from the methyl groups would be expected to increase the basicity of the nitrogen lone-pair. These intrinsic properties ought to be evidenced in the gas phase. Consequently, the order of basicity in the gas phase should be $(CH_3)_3N > (CH_3)_2NH > CH_3NH_2 > NH_3$; experimentally,[11] this is found to be the case. In solution, hydrogen bonding between the solvent molecules and the conjugate acids, R_3NH^+, is also of importance. Hydrogen bonding of the water molecules to the protonated amines should be greatest for NH_4^+; this is because there are more acidic hydrogen atoms in NH_4^+ for the water molecules to bind onto, via n-donation of a lone-pair of electrons in the oxygen atom. Increased solvation of the conjugate acid should enhance the basicity of the corresponding amine. On this basis, the expected order of basicity would be $NH_3 > CH_3NH_2 > (CH_3)_2NH > (CH_3)_3N$. This solution trend is in conflict with the trend in the inherent molecular properties, thus giving rise to the anomalous order of basicity in solution.

Another instance where solvent effects cannot be neglected is exemplified by the basicity of ammonia and aniline. Aniline is a weaker base than ammonia in solution. This is often explained in terms of delocalization of the nitrogen lone-pair in aniline into the aromatic ring, thus rendering the lone-pair less available for donation to a proton. Again such a rationalization invokes the inherent molecular properties and must also operate in the gas phase; however, aniline is a stronger base than ammonia in the gas phase.[7] Presumably, the more effective solvation of NH_4^+ relative to $PhNH_3^+$ is responsible for the solution result. Nevertheless, it should not be concluded that delocalization effects are of no importance in the gas phase. For example, the basicity of aniline is less than that of cyclohexylamine in either solution or the gas phase. This is probably a true reflection of the delocalization of the nitrogen lone-pair in aniline with the aromatic nucleus. A more spectacular effect is afforded by the relative acidities of toluene and water. In the gas phase, ^-OH is sufficiently basic to deprotonate toluene to form the benzyl anion, $PhCH_2^-$;[7] the reverse is true in solution, where $PhCH_2^-$ is some 10^{20} times more basic than ^-OH, as measured by the pK_a values of the conjugate acids. This dramatic alteration in relative base strengths is caused by solvent effects, at least in part. In solution, ^-OH is very much more effectively stabilized by hydrogen bonding than $PhCH_2^-$; in the absence of a solvent, the $PhCH_2^-$ ion is more extensively stabilized, by charge polarization and delocalization, than ^-OH.

5-4 ADDITION AND SUBSTITUTION REACTIONS

Addition and substitution reactions are well known in solution chemistry; analogous processes may be observed in the gas phase.

A. Association Reactions

These processes, leading to the formation of an ion coordinated to several neutral molecules, may be studied conveniently using high-pressure sources or a flowing afterglow apparatus. As an increasingly large number of neutral molecules become attached to the central ion, the gas-phase situation begins to resemble that pertaining in solution in the vicinity of ions. In this way, it is possible to bridge the gap between gas-phase and solution chemistry. This in turn permits a more precise understanding of the nature of solvation of anions and cations by neutral (i.e., solvent) molecules.[12]

For example, the association of water molecules around a central cation or anion may be investigated.[13] The binding energies associated with addition of successive water molecules decrease as more water molecules are already present. Furthermore, the binding energies for the addition of the first few water molecules may be interpreted in terms of a simple charge/radius criterion. Thus, in progressing down a series of homologous ions (for example, Li^+, Na^+, K^+, Rb^+, Cs^+ or F^-, Cl^-, Br^-, I^-), the binding energies decrease as the size of the central ion increases. Moreover, for isoelectronic anions and cations (for example, Na^+ and F^- or K^+ and Cl^-), the binding energies for addition of the first few water molecules are greater for the smaller cation than the larger anion. However, the larger anion can eventually accommodate more water molecules in the first solvation sheath. Consequently, when several water molecules are already present in the cluster ion, it may be more favorable to add more water molecules to a cluster containing a larger anion.

The water molecules presumably bind to the cations by donation of a lone-pair of electrons on the oxygen atom (Fig. 5-3a). However, hydrogen bonding to the anions conceivably could involve either or both hydrogen atoms (Fig. 5-3b). These possibilities create more scope for the binding of water molecules to anions. For halide anions, calculations[14] suggest that the linear hydrogen bond, involving only one of the two hydrogen atoms, is more favorable.

For more complicated neutral species, for instance, acetonitrile or acetone, the hydrogen bonding is more complex and the relatively straightforward trends found for water are not observed.

A particularly important example of cation solvation studies concerns the hydration of a proton. This may be studied by investigating the equilibrium

$$H^+(H_2O)_n + H_2O \rightleftharpoons H^+(H_2O)_{n+1} \tag{5-7}$$

The binding energies for addition of successive water molecules can be determined. Each additional water molecule is more weakly bound than the previous one; a

(a)   (b)   Figure 5-3

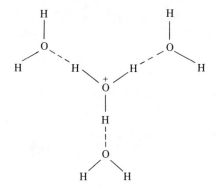

Figure 5-4

gradual decrease in binding energy is observed and there seems to be no special stability associated with the formation of the Eigen structure (Fig. 5-4).

It is also possible to study the behavior of bases in cluster ions. For example, the following equilibrium can be investigated:

$$H_3O^+(H_2O)_n + B \rightleftharpoons BH^+(H_2O)_m + (n-m+1)H_2O \qquad (5\text{-}8)$$

For $m = n = 0$, Eq. (5-8) reduces to a simple proton transfer between H_3O^+ and B. However, as m and n are increased, solvation effects become important in determining the relative basicity of B and H_2O in hydrated H_3O^+. For example, H_2S is a stronger base than H_2O in the absence of other water molecules, but H_2O is a stronger base than H_2S in hydrated H_3O^+.[15]

The examples above serve to illustrate how the boundary between isolated ion and solution behavior becomes blurred when successive neutral molecules are allowed to bind onto the central ion. The investigation of these ion-molecule clusters can furnish insight into the nature of the solvation of ions in solution. In particular, hydrogen bonding may be studied.

B. Substitution Reactions

Bimolecular substitution reactions can be observed using an ICR machine or a flowing afterglow apparatus. A variety of thermochemical data is accessible: e.g., the overall enthalpy change and the forward activation energy. In addition, the bimolecular rate constant may be measured.

Representative rate constants and forward activation energies for bimolecular substitution processes occurring in solution and in the gas phase are given in Table 5-2.[16]

Two trends are immediately apparent from the data of Table 5-2. In the first instance, bimolecular substitutions occur very much more rapidly in the gas phase compared to solution. Displacements involving halogen anions take place some 10^{17} times faster in the absence of solvent than in aqueous media. This reflects the effect of solvation in reducing the nucleophilicity of the incoming anion for reactions occurring in solution. The absolute magnitudes of the rate constants for

Table 5-2 Bimolecular rate constants and apparent activation energies for substitution occurring in solution and in the gas phase

Reaction	Aqueous solution		Gas phase	
	$k_{298\,K}$†	$E_a^{\ddagger}$	$k_{297\,K}$†	$E_a^{\ddagger}$
$CH_3F + {}^-OH \rightarrow CH_3OH + F^-$	9.7×10^{-28}	90.3	$(2.5 \pm 0.5) \times 10^{-11}$	11.3 ± 0.4
$CH_3Cl + {}^-OH \rightarrow CH_3OH + Cl^-$	1.0×10^{-26}	101.6	$(1.5 \pm 0.3) \times 10^{-11}$	1.2 ± 0.6
$CH_3Br + {}^-OH \rightarrow CH_3OH + Br^-$	2.3×10^{-25}	96.1	$(9.9 \pm 0.4) \times 10^{-9}$	2.1 ± 0.6
$CH_3Br + F^- \rightarrow CH_3F + Br^-$	5.6×10^{-28}	105.3	$(1.2 \pm 0.2) \times 10^{-9}$	1.3 ± 0.6
$CH_3Br + Cl^- \rightarrow CH_3Cl + Br^-$	8.2×10^{-27}	103.2	$(2.1 \pm 0.4) \times 10^{-11}$	10.9 ± 0.4

† Values in $cm^3\,molecule^{-1}\,s^{-1}$.
‡ Values in $kJ\,mol^{-1}$.

the gas-phase substitutions reported in Table 5-2 are comparable to the maximum limit (ca. $2 \times 10^{-9}\,cm^3\,molecule^{-1}\,s^{-1}$) derived from a simple collision model (see Sec. 5-1).

The second trend which is evident from Table 5-2 is that although bimolecular displacements have rather large activation energies in solution (approximately $100\,kJ\,mol^{-1}$), the gas phase reactions occur with much smaller activation energies. This indicates that the activation energies found in solution arise from solvent effects instead of from the intrinsic nature of the reaction. Consider the general substitution process

$$Nu^- + RLG \rightarrow [Nu---R---LG]^- \rightarrow NuR + LG^- \qquad (5\text{-}9)$$

Nu^- is the nucleophile and LG is the leaving group. In solution, the solvent molecules can surround and solvate the small nucleophile, substrate, and leaving group much more effectively than the larger intermediate species $[Nu---R---LG]^-$. As this species is formed, solvation energy is lost and potential energy must be supplied to the system. Consequently, the reaction occurs with $[Nu---R---LG]^-$ as a transition state. Clearly, solvents which solvate anions effectively will provide more unfavorable media than those which can only solvate anions to a limited extent. Thus, the increased rates of solution bimolecular processes in dipolar aprotic solvents, compared to protic solvents, may be understood. In the gas phase, no such solvent interactions occur; therefore, most of the activation energy observed in solution is removed, leaving only a very small remnant, which is associated with the inherent nature of the reaction. Moreover, a different potential energy profile is observed (Fig. 5-5).[17] Initial association of the incoming nucleophile with the neutral species occurs; this process is exothermic and leads to an adduct, $[Nu---RLG]^-$, in which the new Nu—R bond is partially formed but the R—LG bond is essentially unbroken. This adduct may isomerize to a second similar species, $[NuR---LG]^-$, in which the new Nu—R bond is essentially formed but the old R—LG bond remains partially unbroken. This second adduct then dissociates to form the products. The species $[Nu---R---LG]^-$, in which both relevant bonds are partially broken, remains

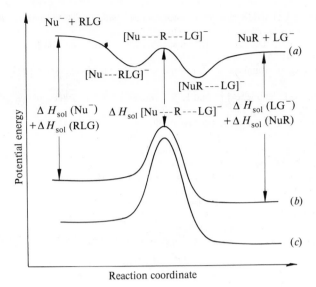

Figure 5-5 Representative potential energy profiles for a nucleophilic displacement occurring (a) in the gas phase, (b) in a dipolar aprotic solvent, and (c) in a protic solvent.

a transition state in the gas phase; however, this transition state is usually lower in energy than either reactants or products.

The existence of two potential energy wells in the energy profile for these displacement reactions may be deduced from several kinds of experimental evidence, including analysis of the absolute rate constants. Perhaps the most pertinent evidence stems from a study of chloride anion transfer from the adduct $[^{35}Cl\text{---}CH_3{}^{37}Cl]^-$ to CH_3CF_3:

$$[^{35}Cl\text{---}CH_3{}^{37}Cl]^- + CH_3CF_3 \begin{array}{l} \nearrow [^{35}ClCH_3CF_3]^- + CH_3{}^{37}Cl \\ \\ \not\searrow [^{37}ClCH_3CF_3]^- + CH_3{}^{35}Cl \end{array} \qquad (5\text{-}10)$$

Only $^{35}Cl^-$ transfer is observed, thus indicating that there is an energy barrier toward interchange of the two Cl^- anions.[18]

REFERENCES

1. For a detailed review, see J. H. Futrell and T. O. Tiernan: In P. Ausloos (ed.), "Fundamental Problems in Radiation Chemistry," Wiley-Interscience, 1968.
2. See, for example, P. F. Knewstubb: "Mass Spectrometry and Ion-Molecule Reactions," chap. 1, Cambridge University Press, 1969.
3. C. Lifshitz and B. G. Reuben: *J. Chem. Phys.*, vol. 45, p. 1925, 1966.
4. For examples of the application of "trapped-ion" ICR see R. T. McIver: *Rev. Sci. Instrum.*, vol. 41, p. 555, 1970; M. T. Bowers, D. H. Aue, H. M. Webb, and R. T. McIver: *J. Am. Chem. Soc.*, vol. 93, p. 4314, 1971.

5. J. L. Beauchamp, L. R. Anders, and J. D. Baldeschwieler: *J. Am. Chem. Soc.*, vol. 89, p. 4569, 1967.

6. J. I. Brauman and L. K. Blair: *J. Am. Chem. Soc.*, vol. 92, p. 5986, 1970.

7. J. F. Wolf, R. H. Staley, I. Koppel, M. Taagepera, R. T. McIver, Jnr., J. L. Beauchamp, and R. W. Taft: *J. Am. Chem. Soc.*, vol. 99, p. 5417, 1977.

8. D. K. Bohme, R. S. Hemsworth, H. W. Rundle, and H. I. Schiff: *J. Chem. Phys.*, vol. 58, p. 3504, 1973.

9. D. B. Dunkin, F. C. Fehsenfeld, A. L. Schmeltekopf, and E. E. Ferguson: *J. Chem. Phys.*, vol. 49, p. 1365, 1968.

10. A. J. Cunningham, J. D. Payzant, and P. Kebarlé: *J. Am. Chem. Soc.*, vol. 94, p. 7627, 1972.

11. J. I. Brauman, J. M. Riveros, and L. K. Blair: *J. Am. Chem. Soc.*, vol. 93, p. 3914, 1971.

12. P. Kebarlé: *Ann. Rev. Phys. Chem.*, vol. 28, p. 445, 1977; see also references cited therein.

13. I. Dzidic and P. Kebarlé: *J. Phys. Chem.*, vol. 74, p. 1466, 1970.

14. G. H. F. Diercksen and W. P. Kraemer: *Chem. Phys. Lett.*, vol. 5, p. 570, 1970; H. Kistenmacher, H. Popkie, and E. Clementi: *J. Chem. Phys.*, vol. 58, p. 5627, 1973.

15. D. K. Bohme, G. I. MacKay, and S. D. Tanner: *J. Am. Chem. Soc.*, vol. 101, p. 3724, 1979.

16. K. Tanaka, G. I. MacKay, J. D. Payzant, and D. K. Bohme: *Can. J. Chem.*, vol. 54, p. 1643, 1976.

17. W. J. Olmstead and J. I. Brauman: *J. Am. Chem. Soc.*, vol. 99, p. 4219, 1977.

18. J. M. Riveros, A. C. Breda, and L. K. Blair: *J. Am. Chem. Soc.*, vol. 96, p. 4030, 1974.

ELECTRON-IMPACT MASS SPECTRA

The subject matter of this chapter has been given in detail in several other texts,[1–5] and hence the present treatment will concentrate on (1) general principles, (2) concise presentation, and (3) the most common classes of compounds.

6-1 MOLECULAR IONS. ISOTOPE ABUNDANCES

The great advantage of mass spectrometry, relative to other analytical techniques which are employed in chemistry, is determination of the integral molecular weight of the compound being investigated if a low-resolution instrument is used, and determination of the molecular formula if a high-resolution instrument is used.

The above generalizations need some qualification, since not all compounds give molecular ions, but in a representative cross section of organic compounds up to molecular weight 600, somewhere around 80 percent may be expected to do so. Molecular ion abundance can be a very useful criterion in assessing the type of compound under investigation. The fraction of ions collected as molecular ions will be roughly proportional to the activation energy for the most facile fragmentation, all other things being equal. Compounds which do not give molecular ions may not do so because (1) the energy of activation for decomposition is very low or zero and no M^+ ions survive to reach the collector, or (2) the sample decomposes thermally prior to ionization. Table 6-1 gives a few loose classifications which may serve as a useful guide to molecular ion abundances in electron-impact mass spectra determined in the range 20–70 eV. "Strong" implies a molecular ion which is the largest peak (base peak) or carries more than, say, 30 percent of the total ion current; "weak" implies a molecular ion of only a few

Table 6-1 Molecular ion abundances in relation to molecular structure

Strong	Medium†	Weak or absent
Aromatic hydrocarbons	Conjugated olefins	Long-chain aliphatic
(ArH)	Ar⌇Br	compounds
ArF	Ar⌇I	Branched alkanes
ArCl	ArCO⌇R	Tertiary aliphatic alcohols
ArCN	ArCH$_2$⌇R	Tertiary aliphatic bromides
ArNH$_2$	ArCH$_2$⌇Cl	Tertiary aliphatic iodides

† In this column, wavy lines indicate a relatively weak bond.

percent of the abundance of the base peak; and "medium" applies to intermediate situations.

While it is possible to say that aromatic hydrocarbons will always give strong molecular ion peaks, and in tertiary alcohols the peaks will be negligible or absent, it is difficult to be precise about behavior between these extremes. For example, in compounds ArCOR or ArCH$_2$R, the molecular ion abundance will decrease quite rapidly with increasing size of the alkyl group R. Lengthening the size of an alkyl chain will reduce the M$^+$ abundance because (1) a larger R group may interact with Ar— or ArCO— to give a new, low-energy fragmentation, and (2) larger, more-branched radicals R tend to be lost more readily because of their greater stability; thus H · is classed as an unstable radical and CH$_3$ ·, C$_2$H$_5$ ·, (CH$_3$)$_2$CH ·, (CH$_3$)$_3$C · have progressively increasing stabilities. These effects usually more than offset the opposing effect which results in less average energy per bond as molecular size increases in a homologous series.

All singly charged ions in the mass spectrum which contain carbon will also

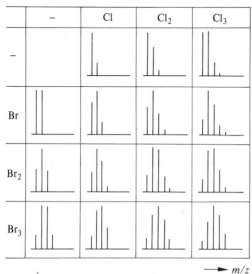

$\longrightarrow$ m/z **Figure 6-1**

give rise to a peak at one mass unit higher. This arises because of the natural abundance of ^{13}C (1.1 percent). For an ion containing n carbon atoms, the abundance of the isotope peak will be $n \times 1.1$ percent of the ^{12}C-containing peak. Thus $C_5H_{12}{}^+$, $C_{40}H_{70}{}^+$, and $C_{100}H_{170}{}^+$ would give isotope peaks at one mass unit higher of approximate abundances 5.5 percent, 44 percent, and 110 percent of the abundance of the ions containing ^{12}C only. These isotope peaks are due to $^{13}C^{12}C_4H_{12}{}^+$, etc. Since most of the compounds which are encountered will contain considerably less than 100 carbon atoms, the ^{13}C isotope peaks are usually appreciably less abundant than the molecular ion. Obviously, the probability of finding two ^{13}C atoms in an ion is very low and $M+2$ peaks are accordingly of very low abundance in most cases, as may be seen by reference to some of the spectra reproduced in this chapter.

Although iodine and fluorine are monoisotopic, chlorine consists of ^{35}Cl and ^{37}Cl in the ratio of approximately 3 to 1 and bromine of ^{79}Br and ^{81}Br in the ratio of approximately 1 to 1. Molecular ions (or fragment ions) containing 1, 2, or 3 chlorine or bromine atoms therefore give rise to the patterns shown in Fig. 6-1.

The isotope patterns to be expected from any combination of elements can readily be calculated, and provide a useful test of ion composition in those cases where polyisotopic elements are involved (see also Sec. 6-6A). Most of the

Table 6-2 Atomic weights and approximate natural abundance of some isotopes

Isotope	Atomic weight ($^{12}C = 12.000000$)	Natural abundance, %
^1H	1.007825	99.985
^2H	2.014102	0.015
^{12}C	12.000000	98.9
^{13}C	13.003354	1.1
^{14}N	14.003074	99.64
^{15}N	15.000108	0.36
^{16}O	15.994915	99.8
^{17}O	16.999133	0.04
^{18}O	17.999160	0.2
^{19}F	18.998405	100
^{28}Si	27.976927	92.2
^{29}Si	28.976491	4.7
^{30}Si	29.973761	3.1
^{31}P	30.973763	100
^{32}S	31.972074	95.0
^{33}S	32.971461	0.76
^{34}S	33.967865	4.2
^{35}Cl	34.968855	75.8
^{37}Cl	36.965896	24.2
^{79}Br	78.918348	50.5
^{81}Br	80.916344	49.5
^{127}I	126.904352	100

remaining elements normally found in organic compounds are essentially mono-isotopic, with the exception of sulfur and silicon (Table 6-2).

Simple chemical reagents of mixed, but known, isotope content may be used to determine the number of given functionalities in a molecule. For example, esterification with a 50:50 mixture of CH_3OH and CD_3OH of a compound containing an unknown number of carboxylic acid groups, followed by observation of the mass spectrum in the molecular ion region, would yield information on the number of ester groupings. For one ester group, the molecular ion would consist of two equally abundant peaks three mass units apart (due to CO_2CH_3 and CO_2CD_3), whilst two ester groups would yield a molecular abundance ion pattern of 1:2:1 at respective masses of M, $M+3$ and $M+6$ where M is the mass of the undeuterated ester. The isotope patterns for larger numbers of functionalities are a matter of simple permutations and mixtures of other isotope reagents may be designed to fit different systems, e.g., the use of $(CH_3CO)_2O/(CD_3CO)_2O$ to determine the number of groups which can be acetylated in a given molecule.

In compounds containing C, H, O only, or C, H, O plus an even number of nitrogen atoms, the molecular weight must be *even*. If a molecule contains C, H, O and an odd number of nitrogen atoms only, then its molecular weight will be *odd*. This generalization applies to all stable even-electron molecules, and a substance such as NO constitutes an exception only because it is a stable radical. The ion at highest mass (except for the ^{13}C isotope peak) will usually be the molecular ion, but in the case of an unknown compound this is an unwarranted assumption because (1) the molecular ion may be absent or (2) the peak at highest mass may be associated with an impurity. The latter possibility may be checked by testing whether the ions at lower mass are related to the supposed molecular ion by the expulsion of plausible neutral particles. For example, differences such as 1, 15, and/or 18 mass units between a supposed molecular ion and the nearest fragment ions encourage the belief that the ion at highest mass is indeed a molecular ion since the loss of H, CH_3, and/or H_2O is relatively common from molecular ions containing C, H, O (although there are, of course, many C, H, O compounds from which these losses do not occur). An extensive list of "common losses" observed from molecular ions in mass spectra is given in Table 6-3; plausible structural inferences for each type of loss are given, but it is emphasized that these are only plausible and should preferably be checked by other techniques.

In addition, many other inferences are possible for the larger masses lost which are listed in Table 6-3; the number of entries is limited only to keep the table fairly concise, and only the more common of the possible inferences are included. The structural information which can securely be inferred from mass spectra alone is frequently limited. Fortunately, the information often complements that available from infrared, ultraviolet, and nuclear magnetic resonance spectroscopy, techniques which should always be used in conjunction with mass spectrometry if possible. In relation to the recognition of molecular ions, it is important to note that the "loss" of 14 mass units should always make one suspect the presence of a homolog, differing in formula by a CH_2 unit. The direct loss of methylene from M^+ is almost never observed since methylene is such a high-energy neutral species.

In compounds containing only C, H, O, N, the loss of 5–13 units is virtually impossible, since loss of many hydrogen atoms (or molecules) would have too high an energy requirement. Loss of 3–5 hydrogens is rarely observed, and usually

Table 6-3 Some common losses from molecular ions

Ion	Groups commonly associated with the mass lost	Possible inference
M − 1	H	—
M − 2	H_2	—
M − 14	—	Homolog?
M − 15	CH_3	—
M − 16	O	$Ar-NO_2$, $\geqslant \overset{+}{N}-\overset{-}{O}$, sulfoxide
M − 16	NH_2	$ArSO_2NH_2$, $-CONH_2$
M − 17	OH	—
M − 17	NH_3	
M − 18	H_2O	Alcohol, aldehyde, ketone, etc.
M − 19	F	⎫
M − 20	HF	⎭ Fluorides
M − 26	C_2H_2	Aromatic hydrocarbon
M − 27	HCN	⎧ Aromatic nitriles ⎩ Nitrogen heterocycles
M − 28	CO	Quinones
M − 28	C_2H_4	⎧ Aromatic ethyl ethers ⎩ Ethyl esters, n-propyl ketones
M − 29	CHO	—
M − 29	C_2H_5	Ethyl ketones, $Ar-n-C_3H_7$
M − 30	C_2H_6	—
M − 30	CH_2O	Aromatic methyl ether
M − 30	NO	$Ar-NO_2$
M − 31	OCH_3	Methyl ester
M − 32	CH_3OH	Methyl ester
M − 32	S	—
M − 33	$H_2O + CH_3$	—
M − 33	HS	⎫
M − 34	H_2S	⎭ Thiols
M − 41	C_3H_5	Propyl ester
M − 42	CH_2CO	⎧ Methyl ketone ⎩ Aromatic acetate, $ArNHCOCH_3$
M − 42	C_3H_6	⎧ n- or isobutyl ketone, ⎩ Aromatic propyl ether, $Ar-n-C_4H_9$
M − 43	C_3H_7	Propyl ketone, $Ar-n-C_4H_9$
M − 43	CH_3CO	Methyl ketone
M − 44	CO_2	⎧ Ester (skel. rearr.) ⎩ Anhydride
M − 44	C_3H_8	—
M − 45	CO_2H	Carboxylic acid
M − 45	OC_2H_5	Ethyl ester
M − 46	C_2H_5OH	Ethyl ester
M − 46	NO_2	$Ar-NO_2$
M − 48	SO	Aromatic sulfoxide
M − 55	C_4H_7	Butyl ester
M − 56	C_4H_8	⎧ $Ar-n-C_5H_{11}$, $ArO-n-C_4H_9$ ⎨ $Ar-iso-C_5H_{11}$, $ArO-iso-C_4H_9$ ⎩ Pentyl ketone
M − 57	C_4H_9	Butyl ketone
M − 57	C_2H_5CO	Ethyl ketone
M − 58	C_4H_{10}	—
M − 60	CH_3COOH	Acetate

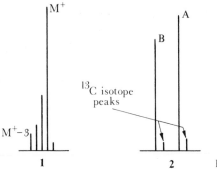

Figure 6-2

occurs when a compound is being dehydrogenated in the inlet system. In such cases, the pattern of ions leading down from M^+ usually shows an ion at each mass (see **1**, Fig. 6-2). The pattern shown in **2** might suggest, for example, that A and B were respectively $M^+ - CH_3$ and $M^+ - H_2O$ ions, since the *specific* loss of 3 hydrogens is not found.

Any structural inferences drawn from Table 6-3 will have the highest probability of being significant when the mass loss in question occurs with a fair degree of specificity (rather than the case where the so-called diagnostic mass loss gives rise to a small peak, surrounded by other larger peaks or peaks of similar size). That Table 6-3 can list only a few possibilities is evident from the consideration that although the loss of 15 mass units is very likely to be loss of CH_3, loss of 57 mass units in a C, H, N, O compound can in theory correspond to loss of C_4H_9, C_3H_5O, C_2HO_2, C_3H_7N, $C_2H_5N_2$, CH_3N_3, C_2H_3NO, or CHN_2O. It will be seen subsequently that the problem of structure elucidation of an unknown is sometimes best approached with the aid of a computer on-line to a high-resolution mass spectrometer (Chap. 9).

6-2 COMMON FRAGMENT IONS

The mass spectrum of an unknown can not only be interpreted by considering the masses of the neutral molecules eliminated from molecular ions but also from a consideration of the masses of fragment ions. Table 6-4 lists the masses of some common fragment ions which may be useful in structure elucidation of simple compounds. Possible inferences should be treated with the kind of caution that was outlined with reference to Table 6-3 in Sec. 6-1.

Inferences drawn from Table 6-4 are most likely to be significant if the peak in question is of great abundance. The table should be used in conjunction with the information given immediately below and in Secs. 6-3 and 6-4, since qualifications not conveniently expressed in the table are necessary in some cases.

Almost no fragmentation reaction can be said to be always characteristic of a certain structural unit, since the so-called characteristic process may be preempted by a lower activation energy reaction upon the introduction of a substituent

Table 6-4 Masses and some possible compositions of common fragment ions

m/z	Groups commonly associated with the mass	Possible inference
15	$CH_3{}^+$	—
18	$H_2O^{\ddagger}$	—
26	$C_2H_2^{\ddagger}$	—
27	$C_2H_3{}^+$	—
28	$CO^{\ddagger}, C_2H_4{}^{\ddagger}, N_2{}^{\ddagger}$	—
29	$CHO^+, C_2H_5{}^+$	—
30	$CH_2{=}\overset{+}{N}H_2$	Primary amine?
31	$CH_2{=}\overset{+}{O}H$	Primary alcohol?
36/38(3:1)	$HCl^{\ddagger}$	—
39	$C_3H_3{}^+$	—
40†	$Argon^{\ddagger}, C_3H_4{}^{\ddagger}$	—
41	$C_3H_5{}^+$	—
42	$C_2H_2O^{\ddagger}, C_3H_6{}^{\ddagger}$	—
43	CH_3CO^+	CH_3COX
43	$C_3H_7{}^+$	C_3H_7X
44	$C_2H_6N^{+}{}_{,}$	Some aliphatic amines
44	$O{=}C{=}\overset{+}{N}H_2$	Primary amides
44	$CO_2{}^{\ddagger}, C_3H_8{}^{\ddagger}$	—
44	$CH_2{=}CH(OH)^{\ddagger}$	Some aldehydes
45	$CH_2{=}\overset{+}{O}CH_3$	⎫ Some ethers and alcohols
	$CH_3CH{=}\overset{+}{O}H$	⎭
47	$CH_2{=}\overset{+}{S}H$	Aliphatic thiol
49/51(3:1)	CH_2Cl^+	—
50	$C_4H_2^{\ddagger}$	Aromatic compound
51	$C_4H_3{}^+$	C_6H_5X
55	$C_4H_7{}^+$	—
56	$C_4H_8^{\ddagger}$	—
57	$C_4H_9{}^+$	C_4H_9X
57	$C_2H_5CO^+$	⎰ Ethyl ketone ⎱ Propionate ester
58	$CH_2{=}C(OH)CH_3^{\ddagger}$	⎰ Some methyl ketones ⎱ Some dialkyl ketones
58	$C_3H_8N^+$	Some aliphatic amines
59	$COOCH_3{}^+$	Methyl ester
59	$CH_2{=}C(OH)NH_2^{\ddagger}$	Some primary amides
59	$C_2H_5CH{=}\overset{+}{O}H$	$C_2H_5CH(OH){-}X$
59	$CH_2{=}\overset{+}{O}{-}C_2H_5$ and isomers	Some ethers
60	$CH_2{=}C(OH)OH^{\ddagger}$	Some carboxylic acids
61	$CH_3CO(OH_2)^+$	$CH_3COOC_nH_{2n+1}$ $(n > 1)$
61	$CH_2CH_2SH^+$	Aliphatic thiol
66	$H_2S_2^{\ddagger}$	Dialkyl disulfide
69	$CF_3{}^+$	—
68	$CH_2CH_2CH_2CN^+$	—
69	$C_5H_9{}^+$	—
70	$C_5H_{10}^{\ddagger}$	—

† Appears as a doublet in the presence of argon from air; useful as a reference point in counting the mass spectrum.

Table 6-4 (*cont.*)

m/z	Groups commonly associated with the mass	*Possible* inference
71	$C_5H_{11}^+$	$C_5H_{11}X$
71	$C_3H_7CO^+$	{ Propyl ketone / Butyrate ester
72	$CH_2=C(OH)C_2H_5\ddagger$	Some ethyl alkyl ketones
72	$C_3H_7CH=\overset{+}{N}H_2$ and isomers	Some amines
73	$C_4H_9O^+$	—
73	$COOC_2H_5^+$	Ethyl ester
73	$(CH_3)_3Si^+$	$(CH_3)_3SiX$
74	$CH_2=C(OH)OCH_3\ddagger$	Some methyl esters
75	$(CH_3)_2\overset{+}{Si}=OH$	$(CH_3)_3SiOX$
75	$C_2H_5CO(OH_2)^+$	$C_2H_5COOC_nH_{2n+1}\ (n>1)$
76	$C_6H_4\ddagger$	{ C_6H_5X / XC_6H_4Y
77	$C_6H_5^+$	C_6H_5X
78	$C_6H_6\ddagger$	C_6H_5X
79	$C_6H_7^+$	C_6H_5X
79/81 (1:1)	Br^+	—
80/82 (1:1)	$HBr\ddagger$	—
80	$C_5H_6N^+$	

| 81 | $C_5H_5O^+$ | |

83/85/87 (9:6:1)	$HCCl_2^+$	$CHCl_3$
85	$C_6H_{13}^+$	$C_6H_{13}X$
85	$C_4H_9CO^+$	C_4H_9COX
85		

| 85 | | |

86	$CH_2=C(OH)C_3H_7\ddagger$	Some propyl alkyl ketones
86	$C_4H_9CH=\overset{+}{N}H_2$ and isomers	Some amines
87	$CH_2=CH-\overset{\overset{+OH}{\|\|}}{C}-OCH_3$	$XCH_2CH_2COOCH_3$
91	$C_7H_7^+$	$C_6H_5CH_2X$
92	$C_7H_8\ddagger$	$C_6H_5CH_2$ alkyl
92	$C_6H_6N^+$	

| 91/93 (3:1) | | n-Alkyl chloride ($\geqslant$ hexyl) |

Table 6-4 (*cont.*)

m/z	Groups commonly associated with the mass	*Possible* inference
93/95 (1:1)	CH_2Br^+	—
94	$C_6H_6O^{+\cdot}$	C_6H_5O-alkyl (alkyl $\neq CH_3$)
94	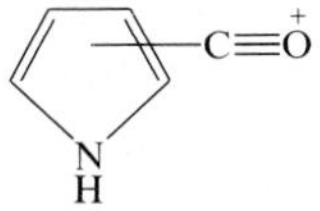 (N-H pyrrole ring)—$C\overset{+}{\equiv}O$	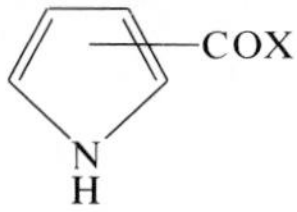 (N-H pyrrole ring)—COX
95	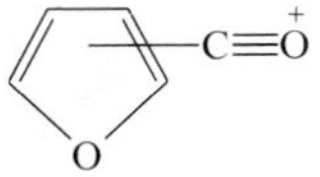 (furan ring)—$C\overset{+}{\equiv}O$	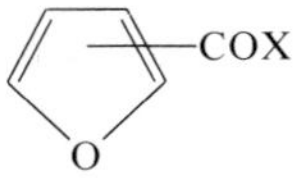 (furan ring)—COX
95	$C_6H_7O^+$	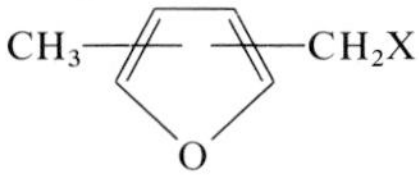 CH_3—(furan ring)—CH_2X
97	$C_5H_5S^+$	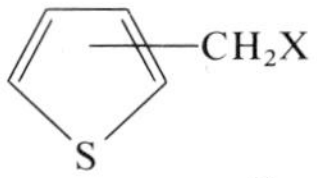 (thiophene ring)—CH_2X
99	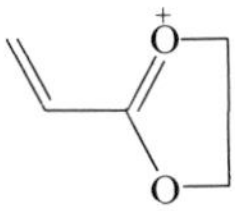 (vinyl-1,3-dioxolane cation)	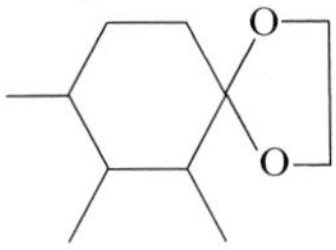 (methyl-substituted cyclohexanone dioxolane ketal)
99	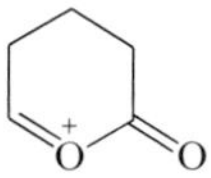 (δ-valerolactone-derived oxocarbenium)	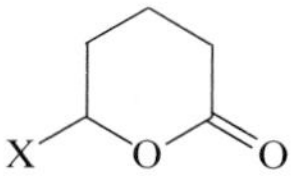 X—(tetrahydropyran-2-one ring)
105	$C_6H_5CO^+$	C_6H_5COX
105	$C_8H_9^+$	CH_3—$C_6H_4CH_2X$
106	$C_7H_8N^+$	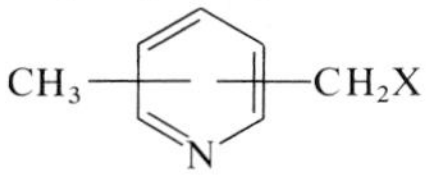 CH_3—(pyridine ring)—CH_2X
107	$C_7H_7O^+$	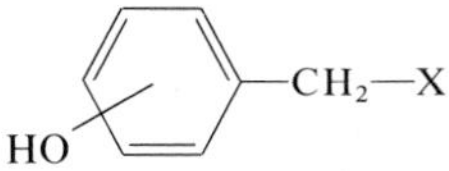 HO—(benzene ring)—CH_2—X
107/109 (1:1)	$C_2H_4Br^+$	—
111	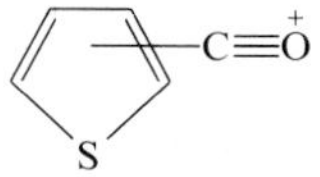 (thiophene ring)—$C\overset{+}{\equiv}O$	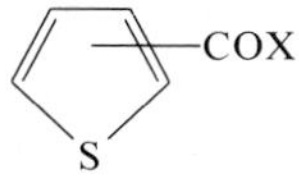 (thiophene ring)—COX
121	$C_8H_9O^+$	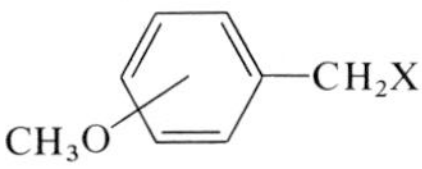 CH_3O—(benzene ring)—CH_2X
122	C_6H_5COOH	Alkyl benzoates
123	$C_6H_5COOH_2^+$	

Table 6-4 (*cont.*)

m/z	Groups commonly associated with the mass	*Possible* inference
127	I$^+$	—
128	HI‡	—
135/137 (1:1)	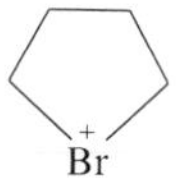	*n*-Alkyl bromide ($\geqslant$ hexyl)
130	C$_9$H$_8$N$^+$	
141	CH$_2$I$^+$	—
147	(CH$_3$)$_2$Si=O—Si(CH$_3$)$_3$	—
149	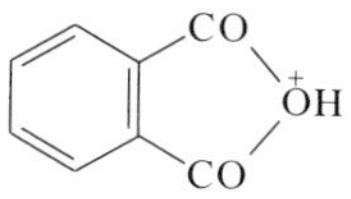	Dialkyl phthalate
160	C$_{10}$H$_{10}$NO$^+$	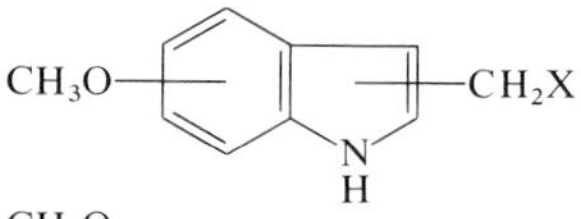
190	C$_{11}$H$_{12}$NO$_2{}^+$	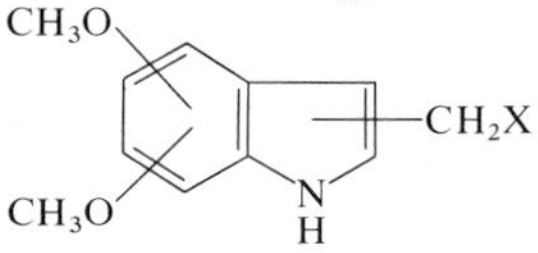

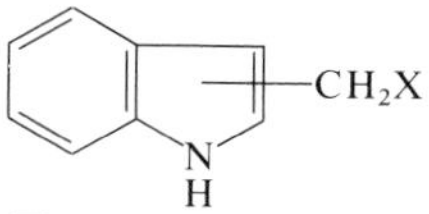

grouping which fragments readily. Therefore, an appreciation of the relative ease of fragmentation of common structural units is extremely important in the interpretation of mass spectra. The relative ease of fragmentation of common aromatic substituents has already been summarized in Table 3-5. Reference to that table shows that loss of a methyl radical to give an abundant M$^+$ − 15 species should occur in virtually all structures ArCOCH$_3$ (where Ar is any aromatic nucleus, substituted or unsubstituted). In contrast, all substituents but fluorine will prevent the so-called characteristic loss of C$_2$H$_2$ from an aromatic ring.

The division of Table 6-4 into areas separated by horizontal dotted lines is useful since mass spectra usually divide into peak groups. For C-, H-, N-, O-containing compounds, the region from m/z 26 to m/z 32 covers C$_2$, CO, CN combinations with varying numbers of hydrogens, depending on the degree of unsaturation of the ion. Similarly, the region m/z 39–45 covers C$_3$, C$_2$O, C$_2$N ions, the region m/z 50–61 covers C$_4$, C$_3$O, C$_2$O$_2$, C$_2$ON, etc., ions, and so on. Ions containing heteroatoms such as fluorine, chlorine, and sulfur often fall outside these characteristic regions.

During the interpretation of a mass spectrum, the interpreter may be doubtful

Table 6-5 Some common impurity peaks

m/z Values	Cause
149, 167, 279	Plasticizers (phthalic acid derivatives)
129, 185, 259, 329	Plasticizer (tri-n-butylacetyl citrate)
133, 207, 281, 355, 429	Silicone grease
99, 155, 211, 266	Plasticizer (tributyl phosphate)
99, 113, 127, 141, ... 309	
285, 299, 313, 327, ... 369	
339, 353, 367, 381	"Hydrocarbon" background
407, 421, 435, 449, ...†	
409, 423, 437, 451, ...†	
64, 96, 128, 160, 192, 224, 256	Sulfur (S_8)
205, 220	2,6-Di-t-butyl-4-methylphenol
446 (and 538)	Polyphenylethers (pump-oils)
45, 59, 73, 87, 89, 101, 103, 117, 133	Polyethyleneglycol
73, 147, 207, 221, 281, 355, 429	SE30 or OV1

† These series of peaks often appear together, the background then appearing as a series of doublets with the components separated by 2 mass units.

as to the significance of a low-abundance ions in the "important" (usually high-mass) region of the spectrum. Are such peaks due to the compound of interest or a contaminant in the sample? Sometimes this question may be answered by preparing a suitable derivative of the compound and searching for the anticipated mass shift. If a derivative is already being examined, then the use of deuteriated reagents (Chap. 11) in the derivatization process is recommended. Additionally, it is important to have on hand a list of ions which arise due to common impurities (Table 6-5).

6-3 ALIPHATIC COMPOUNDS

A. Saturated Hydrocarbons

Saturated normal hydrocarbons undergo cleavage of C—C bonds with a fairly low activation energy. The molecular ion abundance is large in the mass spectrum of ethane, but decreases regularly as the size of the hydrocarbon is increased (Fig. 6-3a). The most abundant fragment ions are members of the series of saturated carbonium ions $C_nH_{2n+1}^+$. The m/z values of these are 29, 43, 57, 71, 85, 99, etc., the most abundant ones being the lower members of the series (Fig. 6-3a). These ions, situated 14 mass units apart, are isobaric with the acyl ions RCO^+ (Sec. 6-3B) so the presence of an abundant m/z 85 ion in a spectrum, in the presence of lower members of the series, suggests the presence of $C_6H_{13}^+$ or $C_4H_9CO^+$ units.

These carbonium ions decompose preferentially by loss of even-electron neutral molecules. The lowest activation energy reactions, which occur with observation of the appropriate metastable transitions, are listed below.

$$C_2H_5^+ \longrightarrow C_2H_3^+ + H_2$$

$$C_3H_7^+ \longrightarrow C_3H_5^+ + H_2$$
$$m/z\ 43 \qquad\qquad m/z\ 41$$

$$C_4H_9^+ \longrightarrow C_3H_5^+ + CH_4$$
$$m/z\ 57 \qquad\qquad m/z\ 41$$

$$C_5H_{11}^+ \longrightarrow C_3H_7^+ + C_2H_4$$
$$m/z\ 71 \qquad\qquad m/z\ 43$$

$$C_6H_{13}^+ \nearrow C_4H_9^+ + C_2H_4 \quad m/z\ 57$$
$$m/z\ 85 \searrow C_3H_7^+ + C_3H_6 \quad m/z\ 43$$

$$C_7H_{15}^+ \longrightarrow C_4H_9^+ + C_3H_6$$
$$m/z\ 99 \qquad\qquad m/z\ 57$$

$$C_8H_{17}^+ \nearrow C_4H_9^+ + C_4H_8 \quad m/z\ 57$$
$$m/z\ 113 \searrow C_5H_{11}^+ + C_3H_6 \quad m/z\ 71$$

$$C_9H_{19}^+ \nearrow C_4H_9^+ + C_5H_{10} \quad m/z\ 57$$
$$m/z\ 127 \searrow C_5H_{11}^+ + C_4H_8 \quad m/z\ 71$$

The effect of branching in a saturated hydrocarbon (1) reduces the molecular ion abundance and (2) causes preferential primary cleavage at the site of branching (Fig. 6-3b). Hence, the site of branching may be located by mass spectrometry.

The mass spectra of acyclic monoolefins are dominated by ions of the general formula $C_nH_{2n-1}^+$, but are often similar even when the double bond is in different positions,[6a] except in those cases where the double bond is tetrasubstituted.[6b]

B. Carbonyl Compounds. The Even-Electron Rule

Carbonyl compounds which contain a γ-hydrogen atom undergo the McLafferty rearrangement (**3 → a**), which is analogous to the photochemical Norrish type II rearrangement in ketones.

This reaction has a relatively low activation energy and even in quite complex

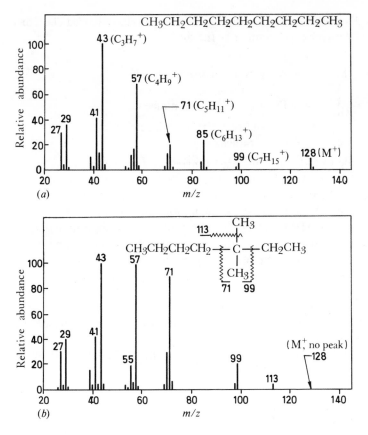

Figure 6-3

molecules is rarely completely "swamped" by other processes. In molecules containing only C, H, (O), ions formed by simple bond rupture without rearrangement of a hydrogen atom or other group will occur at *odd m/z* values. Conversely, fragment ions formed from molecular ions [again containing only C, H, (O)] by bond cleavage with associated rearrangement of a hydrogen atom will occur at *even* masses. Thus in C, H, O compounds, the McLafferty rearrangement ion will always occur at even masses. Ions formed with hydrogen rearrangement are quickly recognized by use of this criterion since examination of the mass spectra of representative C-, H- and C-, H-, O-containing compounds shows that odd *m/z* values are more common than even *m/z* values, especially at lower masses (see Figs. 6-3 and 6-4). This observation is related to the *even-electron rule*, which states that odd-electron ions decompose by loss of radicals or even-electron molecules, whereas even-electron ions decompose by loss of even-electron molecules. Odd-electron ions are those ions which contain an unpaired electron, and include molecular ions ($M^{\ddagger}$) and fragment ions ($A^{\ddagger}$) formed from $M^{\ddagger}$ by loss of an even-electron molecule. Even-electron ions formally do not contain unpaired electrons (possess no radical character). To summarize the rule,

$$M^{\ddagger} \longrightarrow B^+ + \text{radical}$$

or
$$M^{\ddagger} \longrightarrow A^{\ddagger} + \text{even-electron molecule}$$

and
$$A^{\ddagger} \longrightarrow C^+ + \text{radical}$$

or
$$A^{\ddagger} \longrightarrow D^{\ddagger} + \text{even-electron molecule}$$

but
$$E^+ \longrightarrow F^+ + \text{even-electron molecule}$$

and *not*
$$E^+ \longrightarrow G^{\ddagger} + \text{radical}$$

Exceptions to this rule would constitute the occurrence of the last reaction listed, and are surprisingly rare for decompositions which give metastable peaks. Exceptions to the rule are found, for example, in some di-iodides, which successively lose iodine radicals from the molecular ion (another way of looking at the rule is that the successive loss of radicals is "forbidden"). However, note that all the even-electron saturated carbonium ions obey the rule in eliminating even-electron molecules in their lowest energy of activation reactions (Sec. 6-3A). The physical origin of the rule is presumably associated with the greater inherent stability of most even-electron ions relative to ion radicals. Since the elimination of a stable even-electron molecule (H_2O, $CH_2{=}CH_2$, HCN, CH_3COOH, etc.) is usually possible, even-electron ions will undergo such reactions while preserving their even-electron character.

The m/z values of the ionized enols produced by the McLafferty rearrangement in various precursors are listed below. The values are also included in Table 6-4, and we may now add the qualification that the structural inferences require a chain of three saturated carbon atoms adjacent to the carbonyl group and a hydrogen atom in the γ position.

X = H	$m/z\ 44$
X = CH_3	$m/z\ 58$
X = NH_2	$m/z\ 59$
X = OH	$m/z\ 60$
X = C_2H_5	$m/z\ 72$
X = OCH_3	$m/z\ 74$

Note that only in the case of the amide does the ion occur at an odd mass. A methyl substituent on the α-carbon atom will cause a shift in all cases by 14 mass units to higher m/z values. Hence, in *n*-hexanal the McLafferty rearrangement ion occurs at $m/z\ 44$ (Fig. 6-4*a*), whereas in 2-methylpentanal it occurs at $m/z\ 58$ (Fig. 6-4*b*). The McLafferty rearrangement has a low energy of activation and therefore usually dominates low-voltage spectra more than high-voltage spectra.

Dialkyl ketones containing γ-hydrogens in each alkyl chain can undergo consecutive McLafferty rearrangements and also give rise to $m/z\ 58$. For example,

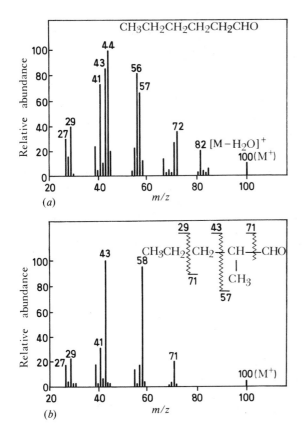

A second favored reaction type in carbonyl compounds is cleavage of the bond adjacent to carbonyl. This is conventionally known as α cleavage and is probably triggered by the unpaired electron on oxygen in the ground state of the ionized carbonyl group ($4 \rightarrow$ **b** and $4 \rightarrow$ **c**, R = alkyl).

$$R-C\overset{+}{\equiv}\overset{\cdot}{O} \xleftarrow{-X\cdot} R\overset{O^{\updownarrow}}{\underset{\mathbf{4}}{C}}X \xrightarrow{-R\cdot} \overset{+}{O}\equiv C-X$$
$$\mathbf{b} \qquad\qquad \mathbf{4} \qquad\qquad \mathbf{c}$$

In ketones (X = alkyl) either X or R may be lost, and in methyl esters both $M^{+}-R$ (COOMe^{+}, m/z 59) and $M^{+}-OMe$ ($M^{+}-31$) ions are usually formed.

Figure 6-4

Primary amides afford both $O^+\equiv C—NH_2$ (m/z 44) and $M^+ - NH_2$ ($M^+ - 16$) ions, but carboxylic acids do not in general lose a hydroxyl radical to a significant extent; carboxylic acids are preferably examined as their derived methyl esters (e.g., prepared by reaction with diazomethane), both from the viewpoint of volatility and interpretation of the resultant mass spectrum.

In unsymmetrical ketones RCOR', where R and R' are different n-alkyl groups, loss of the larger radical usually yields the more abundant fragment ion in 70 eV spectra. However, in low-voltage spectra loss of the smaller radical is usually dominant, indicating that this process has the lower activation energy.

The m/z values corresponding to many of the common acyl ions of types **b** and **c** are listed in Table 6-4. If R is alkyl they are members of the series 29, 43, 57, 71, 85, etc. In general, they decompose further by loss of CO with an appropriate metastable peak to give alkyl ions which fall in the same series of m/z values but displaced to lower masses by 28 units:

$$RCO^+ \xrightarrow[-CO]{m^*} R^+$$

A third diagnostic fragmentation reaction of carbonyl compounds corresponds to γ cleavage:

$$RCH_2 \{\overline{CH_2}CH_2COX\}^{\ddagger}_{\gamma} \longrightarrow CH_2{=}CH{-}\overset{\overset{+}{O}H}{\overset{\|}{C}}{-}X$$
$$X = OMe,\ m/z\ 87$$

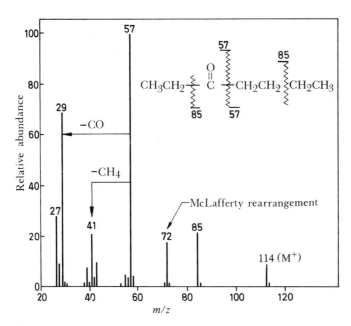

Figure 6-5

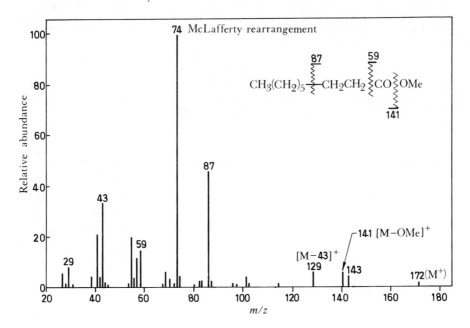

Figure 6-6

Although this ion formally arises by simple bond cleavage, it has been shown[7] by isotopic labeling that cleavage of the γ bond is associated (often to a large extent) with hydrogen rearrangement both to and from the incipient fragment ion. The mechanistic details are not of concern here, but it may suffice to say that the portion of the chain RCH_2 provides a hydrogen for transfer to the carbonyl oxygen, and this is followed by a reciprocal hydrogen transfer from the carbon atom adjacent to the C=O group to the potential neutral species. Several reciprocal hydrogen-transfer reactions are known to occur upon electron impact and can, of course, only be uncovered by deuterium-labeling experiments, since the fragment ion in question appears initially to have arisen by simple bond scission.

γ-Cleavage ions are particularly useful in the mass spectra of methyl esters, where they occur at m/z 87. The mass spectra of an aliphatic ketone and ester are reproduced in Figs. 6-5 and 6-6, to illustrate the abundances of the characteristic cleavage ions. In the spectra of long-chain methyl esters, the formation of fragments of the general formula $(CH_2)_nCOOCH_3^+$, where $n = 2, 6, 10$, etc., is favored. Members of this series fall at m/z 87, 143, 199, 255, 311, 367, etc. The first two members are evident in Fig. 6-6. Notice that although m/z 59 is only of similar abundance to members of the hydrocarbon series 27/29, 41/43, 55/57 in Fig. 6-6, it should be regarded as structurally significant because it appears at a larger m/z value than the saturated C_4 carbonium ion, and therefore must contain oxygen.

The spectra of ethyl and higher esters give only low-abundance molecular ions because a new series of low activation energy reactions is now introduced. The two most general new reactions are loss of the alkyl group of the alcohol with transfer of one or two hydrogens to produce the ionized carboxylic acid and the protonated carboxylic acid, respectively.

$$RCOOR']^{\ddagger} \longrightarrow RCOOH^{\ddagger} + R' - H$$

$$RCOOR']^{\ddagger} \longrightarrow R-C\overset{\displaystyle +OH}{\underset{\displaystyle OH}{\diagdown}} + R' - 2H.$$

In acetates, these ions fall at m/z 60 and 61, and in propionates at m/z 74 and 75. The reactions also occur in aromatic esters and in benzoates ($R=C_6H_5$) afford ions at m/z 122 and 123. The prevalence of double hydrogen transfer (relative to single hydrogen transfer) increases with increasing size of R'. The double hydrogen transfer has a particularly low activation energy and increases in importance in low-voltage spectra.

C. Alcohols, Ethers, and Amines

The ground ionized state of a saturated aliphatic alcohol corresponds to the species **d**, which arises by removal of a nonbonding electron from the oxygen atom of the original alcohol. The ion **d** can in general further decompose to the oxonium ions **e**, **f**, and **g** in the case where R ≠ R' ≠ R''. Where R, R', and R'' are H or n-alkyl groups, the group with the largest number of carbon atoms is expelled to the largest extent in 70 eV spectra.

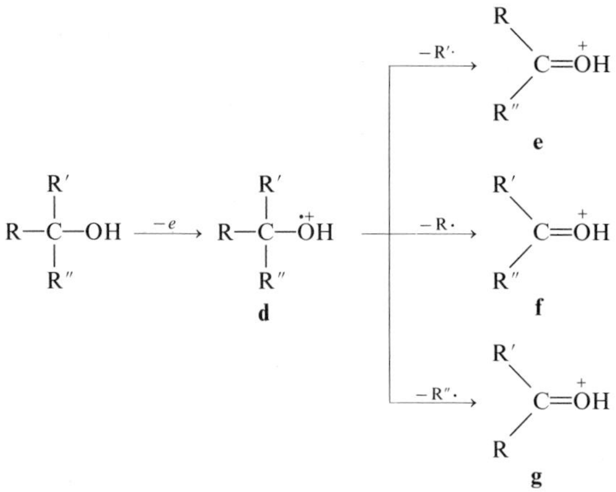

In the case of an n-alkanol the resulting oxonium ion ($CH_2\overset{+}{=}OH$) is found at m/z 31. Depending on the size of the alkyl groups attached to the sp^2-hybridized carbon atom, a series of ions corresponding to saturated oxonium ions will obviously arise at m/z 31, 45, 59, 73, 87, etc. Ions of the same masses may also be produced from saturated aliphatic ether functions (see below). Although ions which are members of the above series may be of moderate or large abundance if

R, R', and R'' are all small [for example, $m/z\,45$ (**h**) is the base peak in the mass spectrum of butan-2-ol (**5**) at 70 eV], they fall in abundance as the length of the alkyl chain(s) is increased.

$$CH_3\!\!-\!\!\underset{\underset{\textbf{5}}{OH}}{CH}\!\!-\!\!C_2H_5 \xrightarrow[-C_2H_5]{-e} \underset{\substack{\textbf{h},\,m/z\,45\\(100\ \text{percent})}}{CH_3CH\!\!=\!\!\overset{+}{O}H}$$

$$\xrightarrow[-CH_3]{-e} \underset{m/z\,59\,(20\ \text{percent})}{C_2H_5CH\!\!=\!\!\overset{+}{O}H}$$

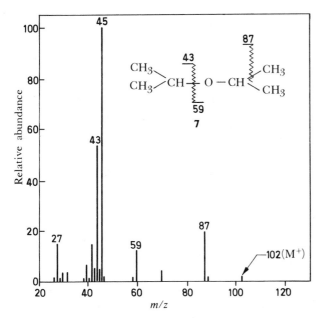

6

In long-chain aliphatic alcohols, the molecular ion is sometimes completely absent and, if not, it is of very low abundance. In the higher alcohols, the first detectable ion is often $M^+ - H_2O$. Unlike thermal loss of water, which is a 1,2-elimination, electron-impact-induced loss of water is a 1,4-elimination (see **6**), as determined by deuterium labeling.

Low-molecular-weight dialkyl ethers undergo primary cleavages similar in nature to those occurring in low-molecular-weight alcohols, viz.:

$$CH_2\!\!=\!\!\overset{+}{O}\!\!-\!\!CH_2R' \xleftarrow{-R\cdot} R\!\!\mid\!\!CH_2\!\!-\!\!\overset{+\cdot}{O}\!\!-\!\!CH_2\!\!\mid\!\!R' \xrightarrow{-R'\cdot} RCH_2\!\!-\!\!\overset{+}{O}\!\!=\!\!CH_2$$

The resulting ions ($m/z\,45, 59, 73, 87,\ldots$) undergo further loss of a neutral olefin if the chain, which was *not* involved in the primary reaction, is ethyl or larger. The eliminated olefin incorporates all the carbon atoms of the alkyl chain in question. Thus, in the case of di-isopropyl ether (**7**, Fig. 6-7), the primary simple

Figure 6-7

bond cleavage can only lead to an $M^+ -15$ ion, which then expels propylene to give **i** (m/z 45).

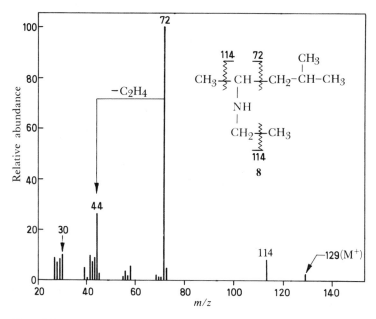

Note that the even-electron rule is obeyed, the even-electron oxonium ion decomposing by loss of an even-electron molecule. In general, the final product ions are still members of the series m/z 45, 59, 73, 87,... Members of this series also arise via C–O cleavage with charge retention by the O-containing portion, but this reaction has a relatively high activation energy and occurs only in ions of high internal energy (see m/z 59 in Fig. 6-7).

The sequence of events occurring in aliphatic amines is analogous to that found in ethers. Initially, a radical will be lost via α cleavage with formation of an immonium ion, and this species then decomposes via olefin elimination when a second chain of two or more carbon atoms is available. Thus in the case of ethyl 1,3-dimethylbutyl amine (**8**, Fig. 6-8), the largest radical is preferentially lost in the

Figure 6-8

70 eV spectrum to give m/z 72, which then eliminates ethylene to give m/z 44 (**j**). To a lesser extent a methyl radical is expelled to give **k** and/or **1** (m/z 114).

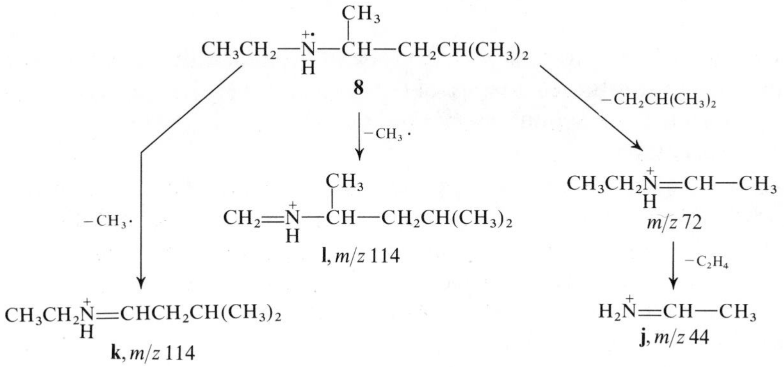

In primary amines the initial radical loss cannot be followed by the type of olefin elimination just described. Simple α cleavage produces the base peak at m/z 30 for compounds RCH_2NH_2:

$$R\overset{\diagdown}{\frown}CH_2\overset{+\cdot}{-}NH_2 \xrightarrow{-R\cdot} \underset{m/z\,30}{CH_2\overset{+}{=}NH_2}$$

All ions which are characteristic of saturated aliphatic amine groupings occur in the series m/z 30, 44, 58, 72, 86, 100, 114, etc. In contrast to the cases of ethers and alcohols, this is an even m/z series.

D. Alkyl Halides

The unique advantage of the mass spectrometer in detecting the presence of Cl and Br due to their characteristic isotopic abundances has already been mentioned (Sec. 6-1). Unfortunately, however, the functional groups F, Cl, Br, and I do not direct the fragmentation strongly. Thus ions of type **m** (X = F, Cl, Br, I), analogous to those found for amines and ethers, do not have a marked stability and are usually absent or of low abundance in the mass spectra of halides.

$$\underset{\textbf{m}}{R'-CH\overset{+}{=}X} \xleftarrow{-R\cdot} \overset{\begin{array}{c}R\\|\end{array}}{R'-CH-X}{\Big]}^{\ddagger} \xrightarrow{-X\cdot} \underset{\textbf{n}}{R'-\overset{+}{C}H-R.}$$

Instead, loss of a halide radical (see **n**) is frequently preferred in the mass spectra of bromides and iodides (reflecting the relative stability of bromine and iodine radicals), while fluorides and chlorides show a more marked tendency to undergo HF and HCl loss.

All n-alkyl chlorides and bromides from hexyl to octadecyl give rise to abundant ions of the formula $C_4H_8X^+$ (m/z 91/93 in chlorides and m/z 135/137 in bromides).[8] The abundance of these ions may be attributed to formation of a five-membered cyclic ion:

$$R\!-\!CH_2 \quad \overset{\cdot\cdot+}{X} \quad CH_2 \quad \xrightarrow{-R\cdot} \quad H_2C \quad \overset{+}{X} \quad CH_2$$
$$H_2C\!-\!CH_2 \qquad\qquad H_2C\!-\!CH_2$$

In the case of branched alkyl chlorides and bromides, the abundance of ions corresponding to substituted analogs of the cyclic ion is greatly reduced, since new low activation energy reactions corresponding to rupture of the chain at the site of branching can compete.

As is general in aliphatic compounds, the abundance of the molecular ion decreases as the chain length and/or chain branching are increased. For a given alkyl halide, the molecular ion abundance increases as the electronegativity of the halogen atom is decreased. However, in the higher *n*-alkyl halide spectra, the molecular ion is often absent or of very low abundance.

E. Relative Fragment-Directing Ability of Common Functional Groups in Aliphatic Compounds

Having indicated in Secs. 6-3A to 6-3D some of the lower activation energy primary and secondary reactions associated with common functional groups, it is natural to enquire whether the fragmentation associated with one functional group will dominate over another when both functionalities are present in an aliphatic molecule. It is almost impossible to make reliable predictions, since any particular arrangement in a complex molecule may lead to some entirely new (and perhaps unique) rearrangement reaction, thus "swamping" the so-called characteristic reactions of the functional groups. However, useful guides are available. In particular, Spiteller and coworkers[9] have estimated the relative powers of common functional groups to initiate cleavage by taking a series of compounds (**9**) and measuring the relative ion yields (i_{OCH_3} and i_X) due to cleavages initiated by

Table 6-6 Relative propensities of common functional groups to produce cleavage in compounds of the general formula 9

X	i_X†	X	i_X†
—COOH	2	—SCH$_3$	90
—CH$_2$OH	6	—OCH$_3$	100
—Cl	8	—NHCOCH$_3$	121
—COOCH$_3$	17	—I	131
—Br	23	$\begin{array}{c}>C<\overset{O-CH_2}{\underset{O-CH_2}{}}\end{array}$‡	514
—OH	25		
—SH	36	—NH$_2$	616
$>C=O$‡	41	—N(CH$_3$)$_2$	1200

† Relative to i_{OCH_3} arbitrarily taken as 100 units.

‡ Data refer to the 3-methoxy-8-ketodecane and 3-methoxy-8-ketodecane ethylene ketal derived from **9**.

the OCH_3 group and the X group; interpretation of the spectra was aided by reference to those of the monosubstituted 3-X-decane and 3-methoxydecane. The results are summarized in Table 6-6, and refer to 70 eV spectra obtained with a cooled ion source. The use of a cooled ion source ensures that direct thermal excitation of the molecules is minimized (see Sec. 6-8) and hence the fragmentation pattern is relatively specific.[9] The quantitative aspects of the results are nevertheless of limited value, and the quotation of specific numbers serves primarily to indicate where major changes in the fragment-controlling abilities of groups are found.

$$C_2H_5CH(CH_2)_4CHC_2H_5$$
$$\underset{CH_3O}{|} \qquad \underset{X}{|}$$

9

As might have been anticipated, groups leading to the most stable onium ions, e.g.,

$$-CH{=}\overset{+}{N}(CH_3)_2 \qquad -CH{=}\overset{+}{N}H_2 \qquad -C\overset{\overset{+}{O}}{\underset{O}{\diagdown}}{\rceil}$$

direct cleavage (α cleavage) the most strongly. Although iodine falls in a part of the table which suggests that it controls fragmentation efficiently, this is not through the formation of stable onium ions but rather through loss of an iodine radical which does not give information as to the site of iodine substitution.

The qualitative trends may be discerned by reference to the mass spectra[9] of 8-chloro-3-methoxydecane (**10**), 8-methylmercapto-3-methoxydecane (**11**), and 3-methoxydecan-8-one ethylene ketal (**12**) (see Figs. 6-9 to 6-11). In the case of the chloro compound (**10**), the spectrum is dominated by α cleavage adjacent to the methoxyl group with loss of the larger radical to give m/z 73. The $M^+ - 29$ ion is associated essentially completely with α cleavage adjacent to the methoxyl group with loss of the smaller (ethyl) radical to give m/z 177/179, which then decomposes further as indicated below:

$$\left[C_2H_5{-}\underset{Cl}{\overset{|}{CH}}{-}(CH_2)_4{-}\underset{OCH_3}{\overset{|}{CH}}{-}C_2H_5 \right]^{\ddag}$$

10

$-C_2H_5\cdot$

m/z 177/179

$*\ \Big|\ -HCl$

m/z 141

$*\ \Big|\ -CH_3OH$

m/z 109

$*\ \Big|\ -C_3H_6$

m/z 67

$$\underset{m/z\ 73}{\overset{CH-C_2H_5}{\underset{+OCH_3}{\|}}}$$

$-CH_3OH$ $*$ $*$ $-C_2H_4$

m/z 41 m/z 45

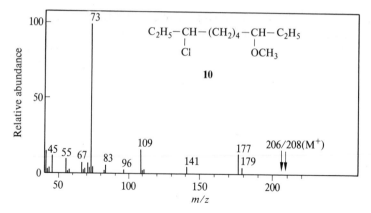

Figure 6-9

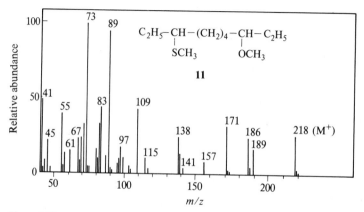

Figure 6-10

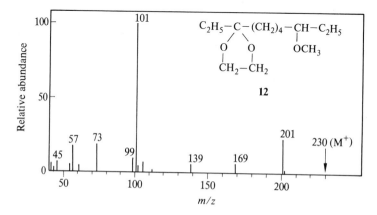

Figure 6-11

In general, where α cleavage can occur at alternative sides of a functional group with loss of radicals of unequal sizes, all other things being equal (e.g., primary radical loss versus primary radical loss, or secondary versus secondary), the loss of the larger radical will be dominant in 70 eV spectra (as in the above case). However in low-voltage spectra (for example, 12 eV) or in the case of metastable transitions, where discernible, the loss of the smaller radical is almost always dominant. Apparently, the loss of the smaller radical occurs with the lower activation energy, but loss of the larger radical becomes the faster reaction in ions of higher internal energy. Obviously, the behavior could be rationalized in terms of a lower frequency factor for the loss of the smaller radical (Sec. 4-1), but such a rationalization does not appear to have a clear physical basis and the phenomenon is not clearly understood at this time.

In the spectrum (Fig. 6-10) of the thioether (**11**), the two possible α-cleavage products of differing mass ($m/z\,73$ and 89) are of similar abundance, with the former decomposing as indicated above and the latter undergoing losses of C_2H_4 and CH_3SH (to $m/z\,61$ and 41, respectively). The losses of C_2H_5, CH_3OH, and SCH_3 from M^+ lead to $m/z\,189$, 186, and 171, with further decomposition as indicated in the reaction scheme:

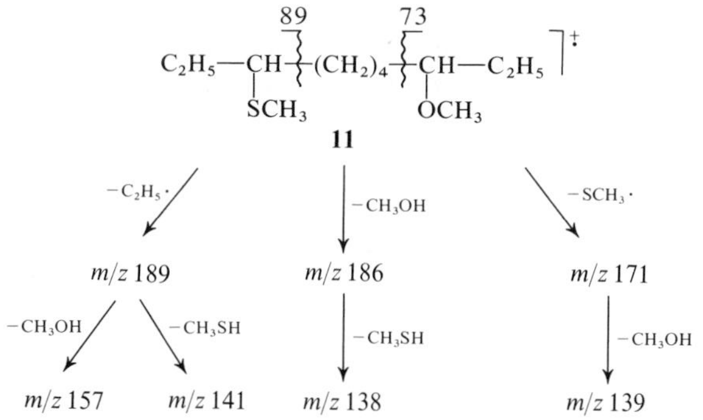

Figure 6-11 needs no comment except to note that the activation energy to form $m/z\,101$ is so low that the ion yield of $m/z\,73$ is greatly reduced relative to the previous cases.

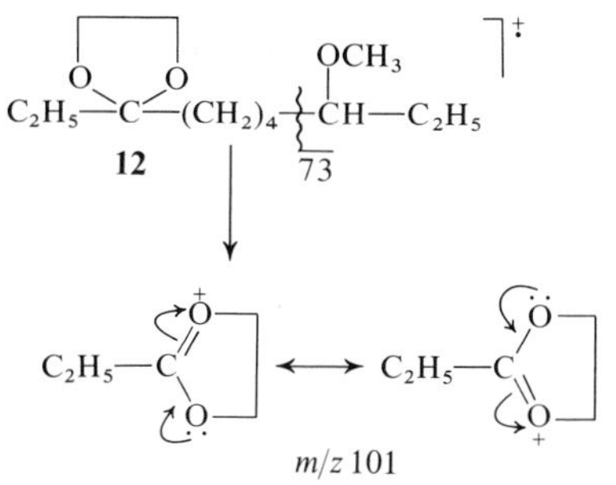

6-4 AROMATIC COMPOUNDS

The major fragmentation paths of compounds C_6H_5X for most common X substituents are given in Table 3-5. In disubstituted aromatics, fragmentation of those substituents X entered near the top of Table 3-5 will in general dominate over, or completely swamp, the cleavages normally associated with X groups entered near the foot of the table.

In some cases the X groups undergo, at 70 eV, characteristic secondary reactions which follow the primary steps already discussed in Chap. 3. These secondary reactions are summarized in Table 6-7.

Table 6-7 Primary and secondary decompositions of some molecular ions C_6H_5X

X	Group expelled in first step	Group expelled in second step
$COCH_3$	CH_3	CO
$C(CH_3)_3$	CH_3	C_2H_4
$COOCH_3$	OCH_3	CO
CHO	H	CO

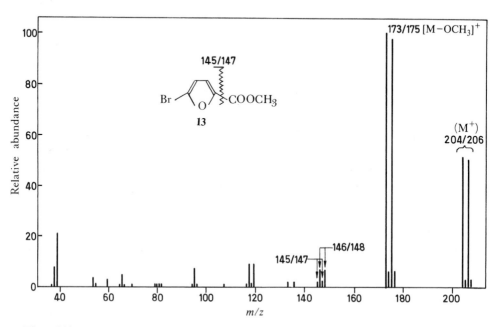

Figure 6-12

The above generalizations dispose of most of the main features of the mass spectra of mono- and disubstituted aromatics for small X groups and *any* aromatic nucleus, except where particularly favorable or unfavorable interactions of substituents and/or heteroatoms occur. This point is demonstrated by reference to the 70 eV mass spectrum of 5-bromo-2-carbomethoxyfuran (**13**, Fig. 6-12), where the anticipated OCH_3 loss completely quenches Br loss as a primary fragmentation.

Complicating interactions, which cause changes from the behavior anticipated on the basis of the spectra of the monosubstituted compounds, may be divided into the two main categories of resonance effects and ortho effects. The former class is easy to understand and even predict. First, those substituents X [$N(CH_3)_2$, $NHCH_3$, NH_2, OCH_3, SCH_3, OH, etc.] which donate electrons to the reaction center Y—Z through a mesomeric effect enhance loss of Z as a radical (see **14 → 15**).

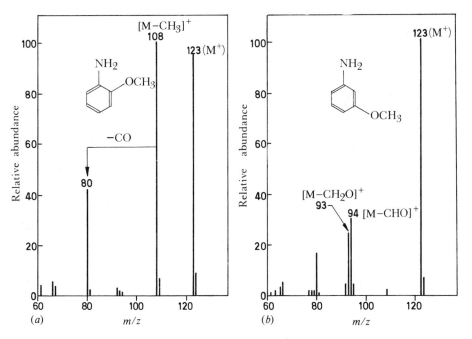

Such effects can operate optimally only when the two substituents are *o-* or *p-* oriented. An example is found in the "normal" loss of CH_2O from the OCH_3 group of *m*-methoxyaniline (Fig. 6-13*b*), in contrast to loss of a CH_3 radical as the

Figure 6-13

lowest activation energy reaction of the molecular ion of *o*-methoxyaniline (Fig. 6-13*a*).

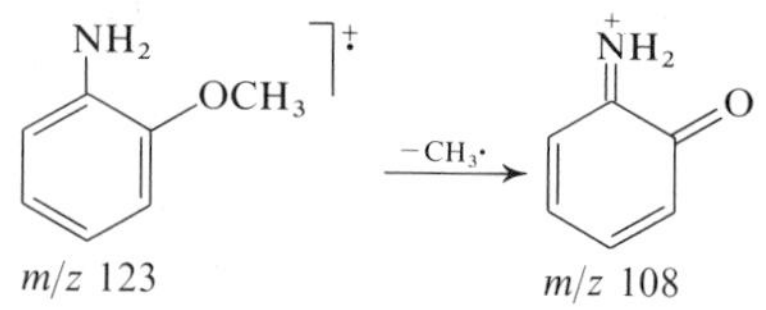

m/z 123 m/z 108

Second, substituents X (NO$_2$, COR, CHO, COOR, etc.) which withdraw electrons from the reaction center Y—Z through a mesomeric effect make the loss of Z as a radical less favorable. For example, the "characteristic" reactions of OCH$_3$, CH$_3$, and NO$_2$ substituents are not grossly different in their energy requirements (AE values of 11.30, 11.80 and 12.16 eV for CH$_2$O, H, and NO$_2$ loss, respectively, in the monosubstituted benzenes). In the absence of interactions between the substituents in events leading up to the transition state, *p*-methoxytoluene would be anticipated to undergo predominantly CH$_2$O loss, and *p*-nitrotoluene hydrogen atom loss. However, since the energy requirements for all the reactions are similar, resonance effects should be considered. It is anticipated that the energy of activation for H loss from *p*-nitrotoluene will be raised [NO$_2$ destabilizes the carbonium ion (**16**)] and for H loss from *p*-methoxytoluene will be lowered (**17**).

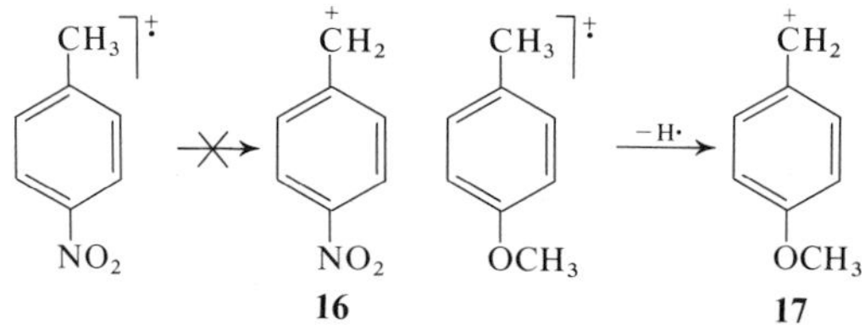

16   **17**

In fact, M$^+$ – H is the most abundant daughter ion arising from primary decomposition of *p*-methoxytoluene (Fig. 6-14) and hydrogen radical loss from *p*-nitrotoluene is negligible relative to NO and NO$_2$ loss (Fig. 6-15).

Similarly, since carbonium ions **18** (for example, X = O, NR) are energetically unfavorable, methyl radical loss from 2-ethylpyridine (**19**) is almost negligible, while from 3-ethylpyridine (**20**) it accounts for the base peak.

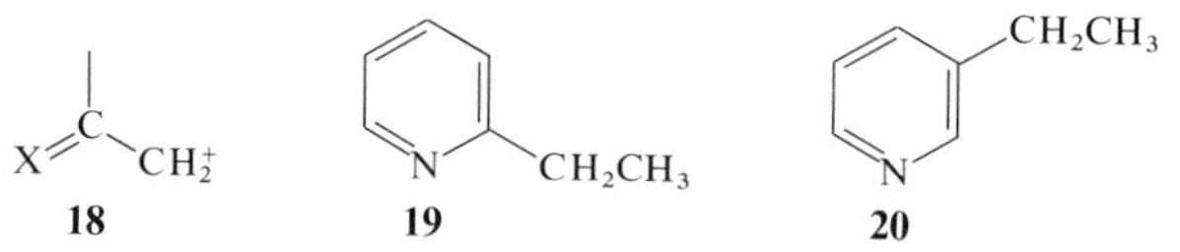

18   **19**   **20**

"Ortho effects" are those in which adjacent substituents interact directly through space, so that a neutral molecule whose constituents originate in part from each substituent may be eliminated with a relatively low activation energy. Thus, the presence of a methyl group ortho to the nitro function in *o*-nitrotoluene introduces a new, low-energy process, namely, loss of a hydroxyl radical (Fig. 6-16;

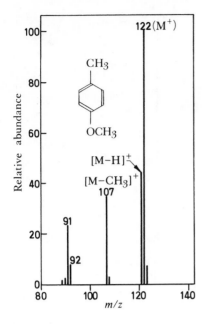

Figure 6-14

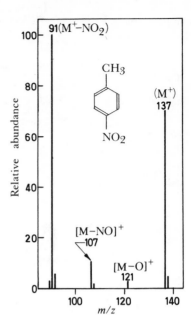

Figure 6-15

contrast Fig. 6-15). One of the most useful ortho effects in structure elucidation is associated with ROH loss (in competition with OR loss) from various kinds of aromatic esters carrying XH groups in the ortho position (page 146).

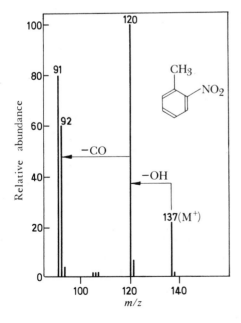

Figure 6-16

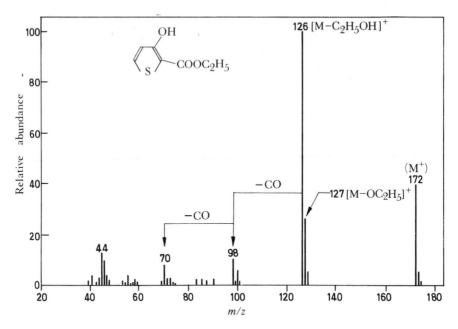

Figure 6-17

$$X = O, S, NH, CH_2$$

M$^+$ − ROH

This ortho effect is illustrated in the mass spectrum (Fig. 6-17) of ethyl-3-hydroxythiophen-2-carboxylate, where the base peak arises due to loss of ethanol from the molecular ion.

6-5 NATURAL PRODUCTS

A. Molecular Formula Determination

The most general use of mass spectrometry in natural products chemistry is for the determination of molecular weight (low resolution) or molecular formula (high resolution). A pre-requisite of such determinations by electron impact is that the intact molecule can be vaporized prior to ionization. Currently, the practical upper limit of molecular weights of natural products that can be determined by mass spectrometry is around 1500, given favorable volatility and thermal stability. Nevertheless, the advantages of molecular formula determinations up to this

Table 6-8 Empirical formula determination of some sterols

	Alcohol A (mp 137°)		Alcohol B (mp 165°)	
Empirical formula	$C_{32}H_{56}O$	$C_{29}H_{50}O$	$C_{32}H_{54}O$	$C_{30}H_{50}O$
Molecular weight	456	414†	454	426†
Carbon content (percent)	84.14	83.99	84.51	84.44
Hydrogen content (percent)	12.36	12.15	11.97	11.81

† Determined by mass spectrometry.

region cannot be overemphasized, especially in the cases of noncrystalline substances available only in microgram quantities, where no alternative approach may be adopted. High-resolution measurements are discussed in detail in Secs. 1-5 and 9-2F.

Frequently, in the case of compounds such as steroids, triterpenes, and alkaloids, microanalysis is insufficient to determine the molecular formula, commonly accepted errors in C and H analyses being up to ± 0.3 percent, which readily accommodates differences in the molecular formula of at least $\pm CH_2$. This point may be illustrated by the example of two sterols (Table 6-8) where mass spectrometry was required to determine the molecular weight.[10]

B. The Shift Rule

A severe limitation of mass spectrometry in structure elucidation is that sometimes there is no simple or predictable relationship between the fragmentation pattern and the structure of the original molecule. This situation may arise if the fragmentation of a similar skeleton has not previously been examined in the mass spectrometer, or if the basic skeleton is so stable to electron impact that only side chains or structurally uninteresting substituents fragment. Conversely, the greatest successes of mass spectrometry have arisen when the basic skeleton (in its natural state, or after suitable derivatization) undergoes such a low-energy fragmentation that this always occurs irrespective of the presence or absence of additional substituents. The first and classical example of the latter type of behavior was observed in the mass spectra of some indole and dihydroindole alkaloids by Biemann and Spiteller,[11] who formulated their observations in terms of a *shift rule*. This rule may be formulated in terms of a molecule AB^+ breaking down into, say, both A^+ and B^+ (for example, in two competing reactions from the molecular ion), and a biogenetically related molecule $R—AB^+$ similarly breaking into $R—A^+$ and B^+. The shift of A^+ to $R—A^+$ indicates the mass, or with high resolution the elemental composition, of the substituent R and its location in a certain part of the alkaloid.

For example, alkaloids possessing the aspidosperma skeleton (**21**) undergo the following sequence of bond ruptures upon electron impact with a relatively low activation energy. In particular, the ion **p** (m/z 124 or a substituted analog) is usually the base peak in these spectra.

21, m/z 282   **o**, m/z 254

q, m/z 152   **p**, m/z 124

Clearly, the primary reactions from **21** are favored since the strain of the polycyclic system is released with simultaneous aromatization of the indole system. The structure assigned to m/z 152 (**q**) is less secure than that assigned to m/z 124 (**p**), but the important point is that sufficient members of the aspidosperma series have been examined to ensure that these ions differ in terms of atomic content only in the additional presence of carbon atoms 3 and 4 (and attached hydrogens) in m/z 152.

The partial mass spectra (showing only peaks of 10 percent or greater relative abundance) of two aspidosperma alkaloids, demethoxypalosine (**22**)[12] and 2,3-dihydrovincadifformine (**23**), are reproduced (Figs. 6-18 and 6-19). The increments in the molecular weights over the basic aspidosperma skeleton are respectively 56 and 58 mass units, which may be associated with C_2H_5CO and $COOCH_3$ substituents through the use of proton magnetic resonance spectroscopy and high-resolution mass measurements. The loss of 28 mass units ($CH_2{=}CH_2$) from M^+ in Fig. 6-18 is replaced by the loss of 86 mass units in Fig. 6-19 ($CH_2{=}CH{-}COOCH_3$). Therefore, the C_2H_5CO substituent of demethoxypalosine (**22**) cannot be attached to either carbon atom 3 or carbon atom 4, but the carbomethoxyl substituent of 2,3-dihydrovincadifformine (**23**) must be located at carbon atom 3 or carbon atom 4. The occurrence of the base peak at m/z 124 (**p**) in both spectra indicates the absence of substituents on atoms 5–10, 20–21 (see **21**), and the occurrence of m/z 152 in Fig. 6-18 indicates further the absence of substituents on carbon atoms 3 and 4 in **22**. The C_2H_5CO substituent of **22** must therefore be located in the indolic part of the molecule, as ascertained through the expenditure of only a few micrograms of sample. Where sufficient material is available, the attachment of the C_2H_5CO group of **22** to the indolic nitrogen may readily be

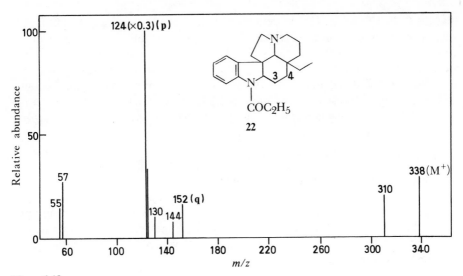

Figure 6-18

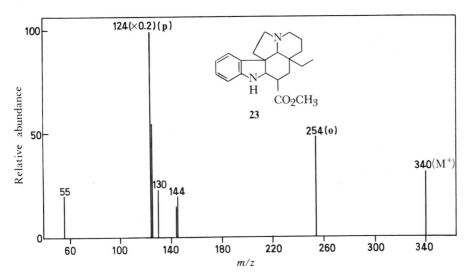

Figure 6-19

established through the application of complementary techniques such as nuclear magnetic resonance, infrared, and ultraviolet spectroscopy.

It is interesting to note that within a spectrum (Fig. 6-18) where the shift rule can be applied to great advantage there occurs also an example of a substituent being lost with a relatively low activation energy *en route* to ions (m/z 130 and 144) which are normally considered to indicate an unsubstituted indole moiety. The ions m/z 130 and m/z 144 are known to contain the indole nucleus plus carbon atom 11, and carbon atoms 10 and 11, respectively (see **o**). It is therefore apparent that the C_2H_5CO substituent of demethoxypalosine (**22**) must be lost as C_2H_4CO with associated hydrogen rearrangement (almost certainly to nitrogen) in the course of the formation of m/z 130 and 144. In contrast, the presence of a methoxyl substituent in the benzene ring of the indole system of such alkaloids *is* uncovered by shifts of m/z 130 and 144 to m/z 160 and 174 respectively. Thus the importance of considering the relative activation energies of various competing reactions before making sweeping generalizations in mass spectrometry is again emphasized.

C. Fragment-Directing Derivatives

In those natural products where the fragmentation pattern upon electron impact is not very specific and is prone to change in an unpredictable manner upon changing the substituent groupings, it may be desirable to convert the compound into a derivative which can strongly control the fragmentation. Suitable derivatives are dimethylamines or ethylene ketals,[2,4] which may be obtained from ketones in the following manner:

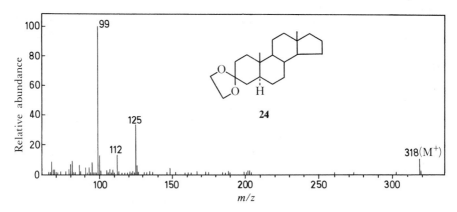

It has been seen (Sec. 6-3E) that such groupings undergo very facile α cleavage. If limited amounts of material are available, it is evident that the one-step preparation of the ethylene ketal from a ketone must be the preferred derivatization. Although the spectrum of 5α-androstan-3-one is very complex, containing large numbers of fragment ions of comparable abundance, the mass spectrum (Fig. 6-20)

Figure 6-20

of the derived ethylene ketal (**24**) is very simple. The origins of the most abundant fragment ions (m/z 99 and 125) can be described in the following terms, as established by replacing the hydrogens thought to participate in hydrogen-transfer reactions by deuterium (Sec. 11-1A):

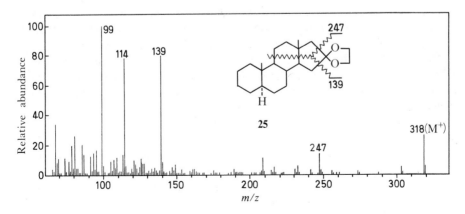

Such reaction sequences are so energetically favorable that analogous processes are found to occur, irrespective of the site of functionalization of the tetracyclic nucleus.[13] For example, the following fragmentations may be anticipated to occur in the mass spectrum (Fig. 6-21) of 5α-androstan-16-one ethylene ketal (**25**), and indeed abundant fragment ions are observed at m/z 139 and 247.

Figure 6-21

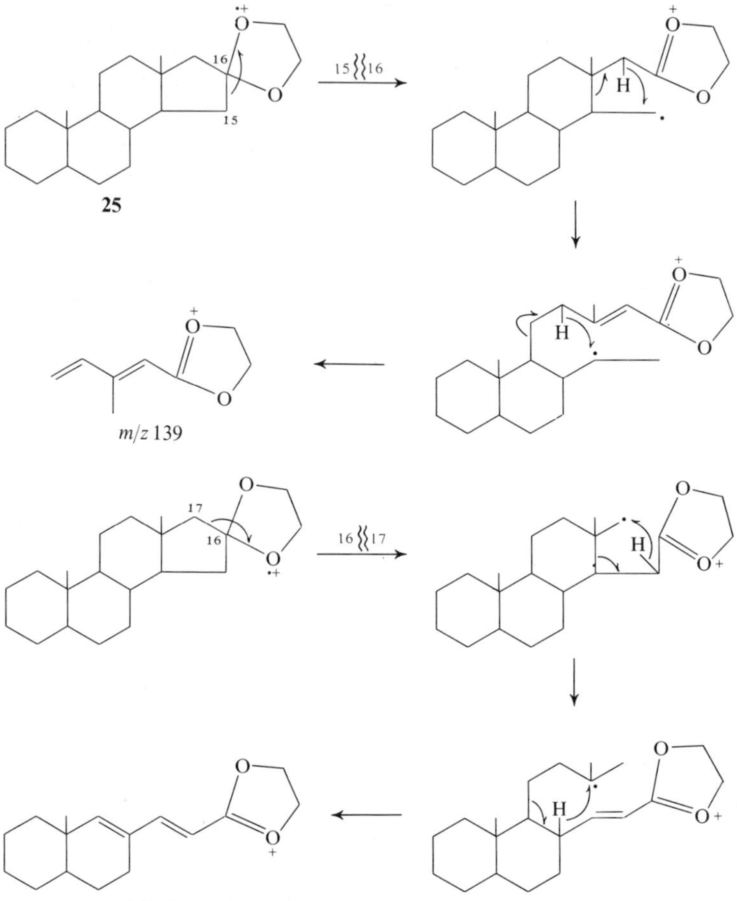

15 ⧸⧸ 16

25

m/z 139

17
16

16 ⧸⧸ 17

m/z 247

Shifts in the *m/z*-values of these conjugated oxonium ions will of course give information as to the nature and position of substituents which have been added to the basic skeleton. An application is to be found in the identification of the substance which controls sexual reproduction in the aquatic fungus *Achlya bisexualis*.[14] This substance, named antheridiol, is active at concentrations as low as 2×10^{-11} g ml^{-1}, and because of the scarcity of material a high-resolution mass spectrum was taken (using a computer and element map output; see Sec. 9-2F) on a few micrograms. The molecular formula was established as $C_{29}H_{42}O_5$. The ions of highest mass in the C—H, C—H—O, and C—H—O$_2$ groups were $C_{19}H_{23}$, $C_{21}H_{30}O$, and $C_{22}H_{30}O_2$, respectively, and the species $C_{19}H_{25}O$ and $C_{19}H_{27}O_2$ were particularly abundant. Such a distribution is strongly suggestive of a steroid containing two oxygen atoms in the tetracyclic system. Assuming the abundant $C_{19}H_{27}O_2$ ion to arise by simple cleavage of the carbon atom 17 and carbon atom 20 bond (see **26** for numbering of the steroid skeleton), the steroid nucleus would have to contain two double bonds. At this point, a steroid-like structure was assumed as a working hypothesis with the remaining ten carbons and three oxygens present as a side chain at carbon atom 17. The position of a third oxygen atom was revealed since the most abundant ion corresponded to $C_{22}H_{32}O_3$, suggesting carbon atom 22 as the point of attachment. The presence of two hydroxyl groups in the hormone was established by a high-resolution mass spectrum of an acetylated product (M^+ of diacetate $C_{33}H_{46}O_7$) and the presence of two double bonds via similar measurements on a hydrogenation product (M^+: $C_{29}H_{46}O_5$). Further considerations of the mass spectra and of infrared (carbonyl bands for α,β-unsaturated ketone and α,β-unsaturated γ-lactone groups) and ^1H NMR spectra, led to the proposal **27** as the most likely structure for the hormone. In particular, the tetrahydro derivative of the hormone afforded an ethylene ketal (**28**) which gave an abundant ion at m/z 141 ($C_7H_9O_3^+$), in accordance with the presence of a 3-hydroxy-7-keto steroid.

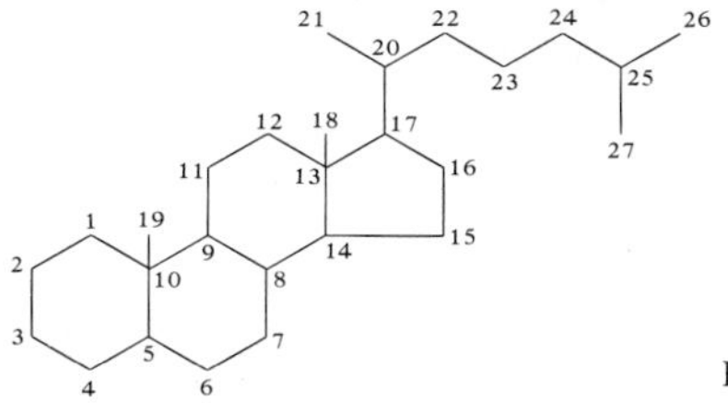

26

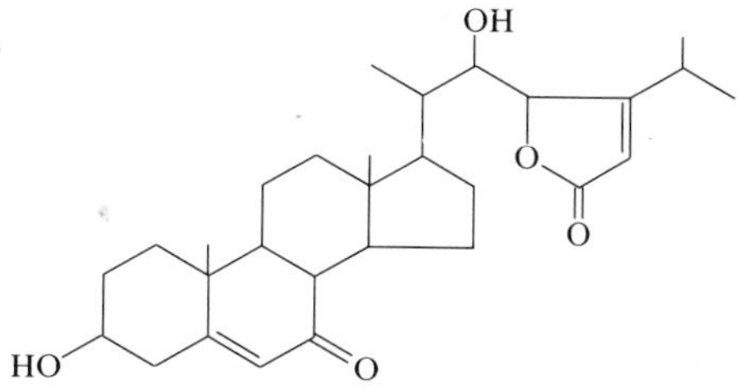

27

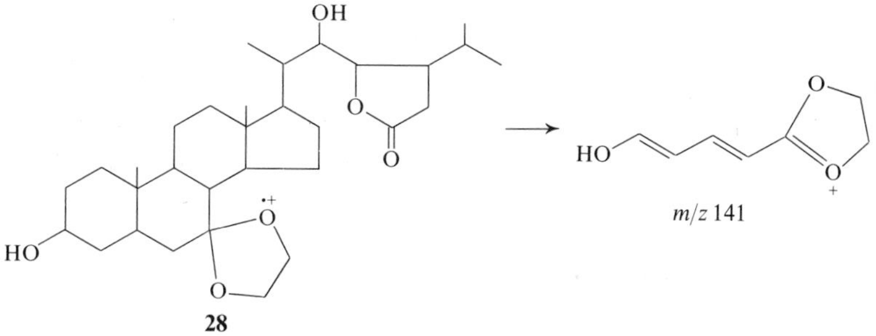

m/z 141

28

The structure **27** has subsequently been confirmed by synthesis.[15] The example is one of many appearing in the literature which attest to the power of high-resolution mass spectrometry. The method is particularly valuable when working with submicrogram quantities. If larger amounts of an unknown are available, then the information provided by the mass spectrometric method is frequently an ideal complement to that provided by ultraviolet, infrared, and NMR spectroscopy.

D. Analysis of Biopolymers

In analyzing the three main types of biopolymers (peptides, nucleotides, and saccharides) by mass spectrometry, a common problem is the preparation of suitable derivatives so that the volatility of substances under investigation is optimized. In addition, only microgram quantities are often available, so mass spectrometry is the analytical method of choice.

a. Peptides Much pioneering work was carried out in this area by Biemann,[1] who showed over a decade ago that sequence information for small peptides could be obtained by mass spectrometry after reduction of the peptide to a polyamino alcohol with lithium aluminum hydride (see also Sec. 10-1C). After much further effort, an important advance was the observation by Lederer and coworkers[16] in 1965 that fortuitine, an acyl nonapeptide methyl ester (**29**), of molecular weight 1359, could be completely sequenced by mass spectrometry.

$$
\underset{\textbf{29}}{CH_3(CH_2)_{18,20}CO-Val-MeLeu-Val-Val-MeLeu-\overset{\overset{Ac}{|}}{Thr}-\overset{\overset{Ac}{|}}{Thr}-Ala-Pro-OMe}
$$

$$
\underset{\textbf{30}}{\left[etc.-N-\overset{\overset{R_n}{|}}{CH}-CO \Big\} N-\overset{\overset{R_{n+1}}{|}}{CH}-CO \Big\} N-\overset{\overset{R_{n+2}}{|}}{CH}-CO-etc. \right]^{\ddagger}}
$$

The sequence analysis was possible because a relatively large fraction of the total ion current was carried by ions formed by cleavage of each peptide bond (see **30**) with charge retention in each case by the left-hand portion of the molecule, i.e., by the portion containing the N terminus. Ions formed by such cleavages are known as sequence ions, and it will be obvious that the mass difference between the consecutive nth and $(n+1)$th sequence ions will be characteristic of the amino acid in the $(n+1)$th position from the N terminus. Extraction of the sequence information was facilitated by the fact that all the useful sequencing ions (which necessarily contain the terminal N-acyl group) could be recognized because they occurred as "doublets" situated 28 mass units apart, since the acyl group contained 18 and 20 methylene groups to similar extents, i.e., the peptide was a mixture of two components.

The most useful subsequent findings largely constitute improvements to the same approach. An important milestone, again from Lederer's laboratory,[17] was the suggestion that peptides should be permethylated prior to examination in the mass spectrometer. Since the —CO—NH— group is converted to a —CO—N(CH$_3$)— group by this procedure, intermolecular hydrogen bonding between amide linkages of adjacent chains is eliminated and volatility improved. Moreover, cleavage at the peptide bonds is now enhanced.

If the mass spectrometric method is to complement classical methods of protein sequencing, it must be possible to take the oligopeptides produced by enzymic cleavage of proteins, and derivatize these in high yield using quantities of only 0.05 to 1.0 micromoles (0.05 to 1.0 mg for a molecular weight of 1000). The best derivatization sequence appears to be acetylation followed by permethylation. The acetylation[18] is carried out by means of acetic anhydride in methanol (1:4) at room temperature for 3 h, and then terminated by evaporation of the reagents by means of a vacuum pump. As a consequence, basic free amino groups in the peptide are acetylated (e.g., the N terminus of the peptide and the basic amino group in the side chain of lysine); hydroxyl groups (e.g., of tyrosine, serine, or threonine) are not acetylated under these conditions. Two methods of permethylation have been employed: (1) the N-acyl peptide derivative in dimethylformamide is treated with an excess of methyl iodide in the presence of silver oxide or (2) the N-acyl peptide derivative is treated with an excess of dimethyl sulfoxide anion ($^-CH_2SOCH_3$) in DMSO, followed by addition of excess methyl iodide; the base $^-CH_2SOCH_3$ is generated by reaction of sodium hydride with DMSO. Both procedures work quite well, but the latter is to be preferred when working with small quantities since adsorption of material on the surface of the silver oxide is potentially a problem; in addition, the silver oxide procedure can lead to extensive C-methylation of glycine and aspartic acid.[19] The permethylation procedure results in the following conversions:

$$-CONH- \longrightarrow -CON(CH_3)-$$

$$-CONH_2 \longrightarrow -CON(CH_3)_2$$

$$\overset{\mid}{-CH(OH)} \longrightarrow \overset{\mid}{-CH(OCH_3)}$$

$$-COOH \longrightarrow -COOCH_3$$

The method of permethylation in dimethyl sulfoxide (DMSO), which is also important in the derivatization of sugars, is of sufficient general importance that experimental details are provided.

A base is prepared by heating sodium hydride in dimethyl sulfoxide (50 mg ml^{-1}) at 90°C for 15 min or until the evolution of H$_2$ ceases. An appropriate quantity of the base (slight excess) is added, when cool, to the acetylated peptide dissolved in dimethyl sulfoxide (0.1 ml), followed by addition of methyl iodide (very large excess). The reaction is allowed to proceed for 1 min[20] in a stoppered tube, after which the permethylated product is isolated by dilution with water and extracted with chloroform (1 ml). The chloroform layer is washed twice with water and evaporated in vacuo.

Three points are noteworthy about the above procedure:

1. Two successive derivatization steps (acetylation and permethylation) may be carried out in the same tube, thereby avoiding losses due to transfer.
2. If the permethylation reaction is not quenched after 1 min, but rather is allowed to proceed for periods of up to 1 h, the sulfur-containing amino acids and histidine are methylated to the stage of forming involatile onium salts, and sequence information will be lost.

3. The dimethyl sulfoxide used for the base generation ($^-CH_2SOCH_3$) and as a solvent must be rigorously dried by periodic distillation from calcium hydride under reduced pressure.

To transfer the sample so obtained into the mass spectrometer, it is dissolved in 20 µl of chloroform, drawn into a fine capillary, and slowly dried onto the quartz tip of the direct insertion probe.

The above permethylation procedure is not successful with arginine. If a sample is known, or suspected, to contain this amino acid, it is usual to subject it to mild hydrazinolysis (treatment with hydrazine hydrate at 75°C for 15 min). This treatment causes cleavage of the guanidine group without significant cleavage of the backbone amide bonds of the peptide. The resulting ornithine residue is then modified by acetylation and permethylation in the normal manner, and after derivatization is characterized by an in-chain mass of 184 mass units.

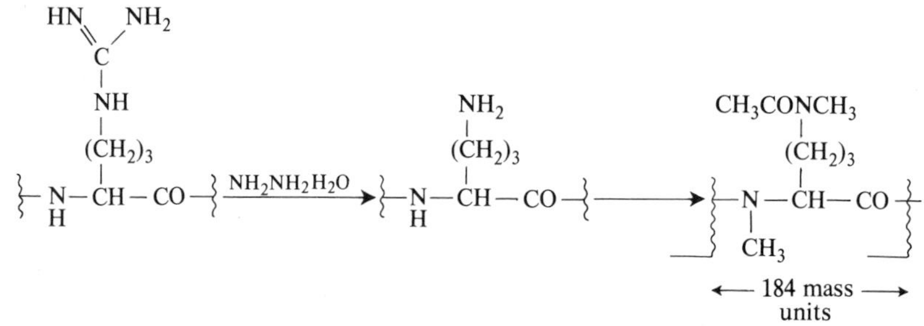

If the peptide (or mixture of peptides), treated in the above manner, contains asparagine or glutamine, then the amide side chain of such residues will be modified as follows during derivatization, e.g., for glutamine:

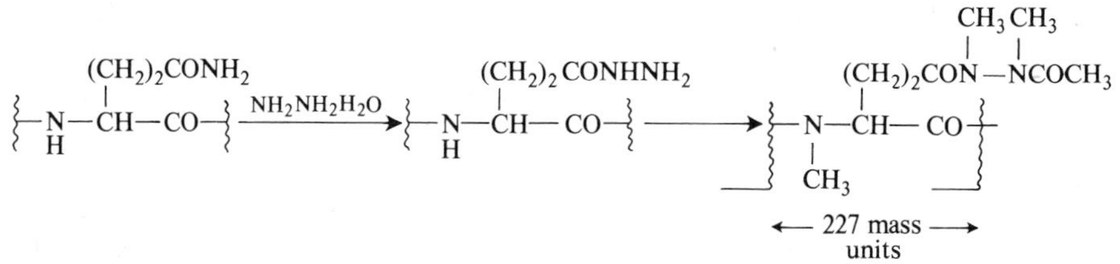

The structures of the derivatized amino acids, following acetylation and permethylation, are given in Table 6-9, together with relevant mass numbers and associated ions. The amino acid cysteine is not included in the table, since, following permethylation and volatilization of the resulting derivative in the mass spectrometer, it sequences as dehydroalanine:

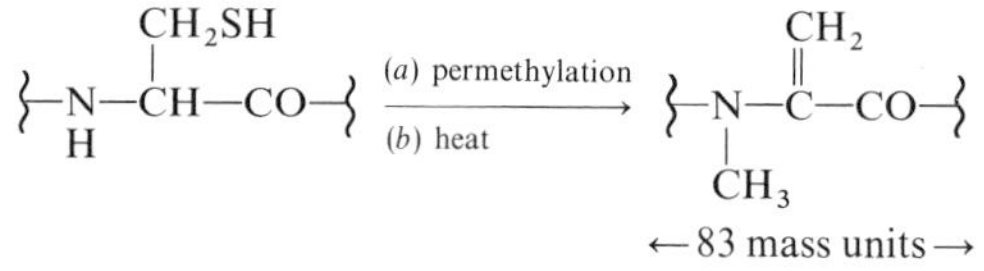

Table 6-9 Integral mass numbers corresponding to derivatized amino acids

Amino acid	Structure of derivative	A N-termi- nal mass	B Mass of residue	C C-termi- nal mass	Associated ions† (commonly observed)
Glycine (Gly)	CH$_3$ | —N—CH$_2$CO—	114	71	102	
Alanine (Ala)	CH$_3$ | —N—CHCO— | CH$_3$	128	85	116	Loss of —CO from A or B
Valine (Val)	CH$_3$ | —N—CHCO— | CH(CH$_3$)$_2$	156	113	144	Loss of —CO from A or B
Leucine‡ (Leu)	CH$_3$ | —N—CH—CO— | CH$_2$ | CH(CH$_3$)$_2$	170	127	158	Loss of —CO from A or B plus further loss of ketene from A (m/e 100)
Serine (Ser)	CH$_3$ | —N—CHCO— | CH$_2$OCH$_3$	158	115	146	Loss of CH$_3$OH from A or B (increases with increasing temperature)
Threonine (Thr)	CH$_3$ | —N—CHCO— | CHOCH$_3$ / CH$_3$	172	129	160	Loss of CH$_3$OH from A or B (increases with increasing temperature); occasional loss of side chain minus H (58 mass units) from B or C
Aspartic Acid (Asp)	CH$_3$ | —N—CHCO— | CH$_2$COOCH$_3$.	186	143	174	Loss of —CO from A or or B
Glutamic Acid (Glu)	CH$_3$ | —N—CHCO— | CH$_2$CH$_2$COOCH$_3$	200	157	188	Loss of —CO from A or B
Asparagine (Asn)	CH$_3$ | —N—CHCO— | CH$_2$CON(CH$_3$)$_2$	199	156	187	Loss of —CO from A or B
Glutamine (Gln)	CH$_3$ | —N—CHCO— | CH$_2$CH$_2$CON(CH$_3$)$_2$	213	170	201	Loss of —CO from A or B

Table 6-9 (*cont.*)

Amino acid	Structure of derivative	A N-terminal mass	B Mass of residue	C C-terminal mass	Associated ions† (commonly observed)
Phenylalanine (Phe)	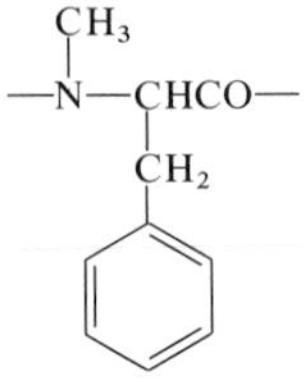 CH₃ / —N—CHCO— / CH₂ / (phenyl)	204	161	192	Loss of —CO from A <u>m/z 91, 162</u>
Tyrosine (Tyr)	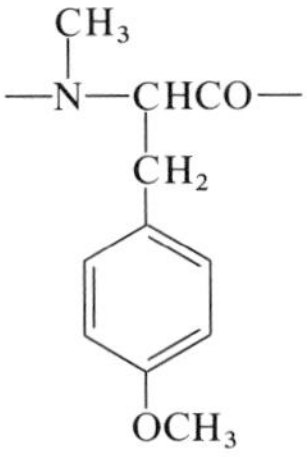 CH₃ / —N—CHCO— / CH₂ / (phenyl) / OCH₃	234	191	222	Loss of —CO from A <u>m/z 121, 161, 192</u>
Tryptophan (Trp)	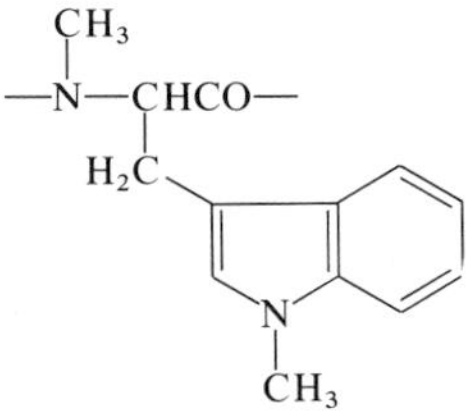 CH₃ / —N—CHCO— / H₂C—(indole) / CH₃	257	214	245	Loss of —CO from A <u>m/z 144, 215</u>
Lysine (Lys)	CH₃ / —N—CHCO— / (CH₂)₄N(CH₃)COCH₃	241	198	229	Loss of —CO from A 171 <u>m/z 171</u>
Histidine (His)	CH₃ / —N—CH—CO— / CH₂-Imidazole (CH₃)	208	165	196	m/z 95; ± 14 m.u. satellites of B.
Ornithine § (Orn)	CH₃ / —N—CHCO— / (CH₂)₃N(CH₃)COCH₃	227	184	215	Loss of —CO from A
Proline (Pro)	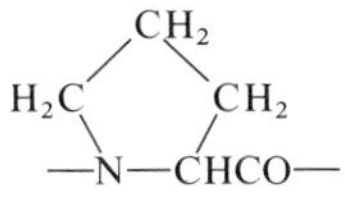 CH₂ / H₂C CH₂ / —N—CHCO—	140	97	128	

† Associated ions which are normally abundant indicated by underlining.
‡ Not differentiated from isoleucine.
§ Result of hydrazinolysis of arginine.

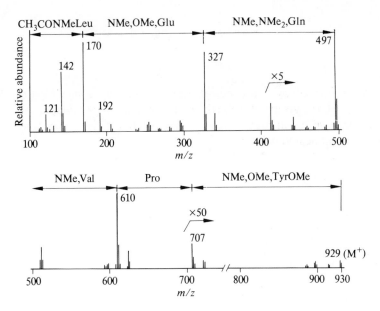

Figure 6-22

The sequencing of a hexapeptide, produced by enzymic hydrolysis of a protein,[21] may be illustrated by reference to Fig. 6-22 and Table 6-9. The abundant ions at m/z 170 and 142 indicate N-terminal leucine (although employing this mass spectrometric method leucine cannot be differentiated from isoleucine). Following m/z 170, the next most abundant ion in the spectrum is at m/z 327 and this corresponds to a Leu . Glu sequence. Further sequence ions of decreasing abundance follow at m/z 497, 610, and 707, giving the mass differences 170, 113, and 97, and so the sequence can be further extended to Leu . Glu . Gln . Val . Pro. The relatively weak satellite ions lying 14 mass units to the high mass side of the sequence ions m/z 327, 497, 610, and 707 correspond to a minor amount of C-methylation. The high relative abundance of the sequence ions is noteworthy; it is emphasized that even if nonsequence ions were of comparable abundance to sequence ions in a given region of the spectrum, the interpretation of the spectrum would not necessarily be complicated because only rarely would extraneous processes lead to mass differences corresponding to an amino acid which was not actually present.

The molecular ion in the spectrum can be identified (with its associated ^{13}C isotope peak) as the highest mass ion in the spectrum, given also the identification aid that normally peaks will appear at 15 and 31 mass units less (loss of CH_3 and OCH_3 from M^+). In the present instance (Fig. 6-22), the C-terminal amino acid is given by the difference of 222 (929 – 707), thus defining the complete sequence as Leu . Glu . Gln . Val . Pro . Tyr. The presence of tyrosine is further confirmed by the occurrence of an abundant ion at m/z 121 (**r**), and its C-terminal position is also indicated by the ion at m/z 192.

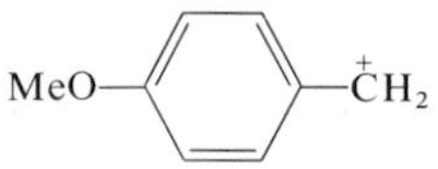

r, m/z 121 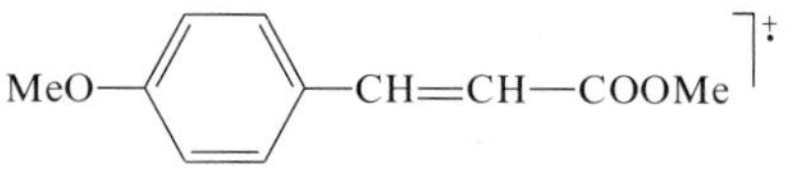 m/z 192

The ion m/z 192 is formed by a process of general importance in which a hydrogen atom of a β—CH_2 group of aspartic acid, asparagine, phenylalanine, histidine, tyrosine, or tryptophan rearranges to the carbonyl group of the preceding amino acid, as indicated below:

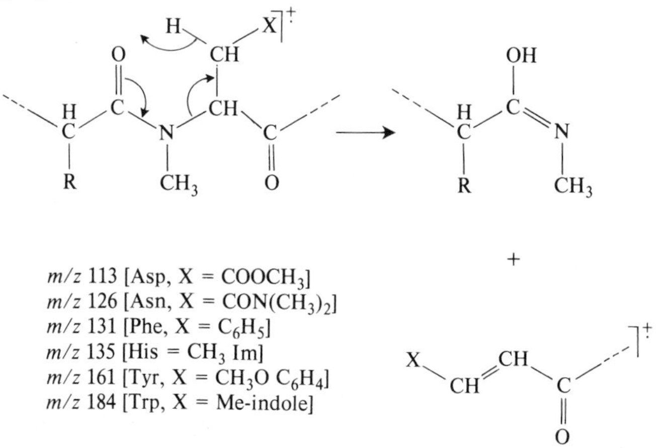

m/z 113 [Asp, X = $COOCH_3$]
m/z 126 [Asn, X = $CON(CH_3)_2$]
m/z 131 [Phe, X = C_6H_5]
m/z 135 [His = CH_3 Im]
m/z 161 [Tyr, X = CH_3O C_6H_4]
m/z 184 [Trp, X = Me-indole]

s

Most frequently, the above-named amino acids are followed by a sequence of several amino acids toward the C terminus of the peptide; the ionic fragment **s** contains these amino acids. Fortunately, **s** fragments by breaking at amide bonds; the charge is retained by the X-containing portion (see **s**), in close analogy to the formation of normal sequence ions. Thus, the m/z values listed for the six amino acids above provide starting points for the extraction of sequence information. For example, the series of ions m/z 126, 253, 368, 559 would indicate that the peptide contained the sequence of amino acids $---$ Asn . Leu . Ser . Tyr $---$. The fragment is said to arise by in-chain fragmentation at Asn.

The fragmentations described thus far for peptides have been used to solve many important problems. For example, in-chain fragmentation at Trp in the mass spectrum of derivatized adipokinetic hormone produced the ion series m/z 184, 255, 384, 428.[22] The last mass difference of 44 mass units corresponds to a C-terminal $N(CH_3)_2$ group, and the series indicated the sequence $---$ Trp . Gly . Thr . NH_2 (i.e., threonine terminating as a primary amide before derivatization). Mass spectrometry was the main method used to establish the sequence pyrrolidone carboxylic acid . Leu . Asn . Phe . Thr . Pro . Asn . Trp . Gly . Thr . NH_2 for adipokinetic hormone, the first peptide hormone from an insect neuro-endocrine organ to be fully characterized. Peptides blocked at the N terminus by pyrrolidone carboxylic acid are readily recognized by mass spectrometry since this residue gives an abundant m/z 98 ion, and usually one of much lesser abundance at m/z 126.

This information is of general utility since the cyclized N terminus may arise on occasions from N-terminal glutamyl residues (as indicated above) before the mass spectrum of a derivative is obtained.

If the peptides being examined are products of enzymic digests, then those peptides which are produced by cleavage within the chain of the original protein will, in many cases, carry only specific amino acids at the C-terminus. Hence identification of M^+ is aided in tryptic digests, since the enzyme trypsin cleaves only after lysine and arginine (i.e., at the C-terminal side of Lys and Arg), while the enzyme chymotrypsin cleaves only after hydrophobic residues (e.g., at the C-terminal side of phenylalanine, tyrosine, and tryptophan; compare Fig. 6-22).

In obtaining spectra such as that shown in Fig. 6-22, it is advantageous to move the tip of the direct insertion probe into the ionization chamber by fractional movements. In this way, a temperature gradient may be obtained at the probe tip, and traces of nonpeptide impurities (which are almost inevitably present in material isolated in small amounts after the use of column chromatography, paper chromatography, etc.) preferentially volatilize first. The progress of this fractionation may be followed by displaying fast scans of the spectrum on an oscilloscope. Since such impurities are usually of a molecular weight considerably less than penta- to decapeptides, it is frequently easier to obtain "clean" spectra of such peptides than of di- to tetrapeptides.

Fractionation of mixed peptides on the probe tip[21,23] can lead to sequence information being obtained from enzymic digests of proteins without the tedious isolation of pure peptides. The procedure[21] is to pass the enzymic digest through a Sephadex column which gives a separation according to molecular weight of peptides present. For mass spectrometric purposes, fractions containing mixtures of peptides of molecular weight less than 1000 are derivatized. For example, a derivatized Sephadex fraction of a tryptic digest of a protein yielded spectra which changed continuously from 150 to 250°C, finally giving a spectrum which appeared to be that of a single component peptide at 260°C. Consecutive scans obtained about the middle of the temperature gradient are reproduced in Fig. 6-23a,b,† the former at a source temperature of 200°C and the latter at a temperature of 220°C. Interpretation is effected by comparing consecutive scans

† In order to simplify the diagrams, the normal amino acid abbreviations are used to represent the derivatized amino acids.

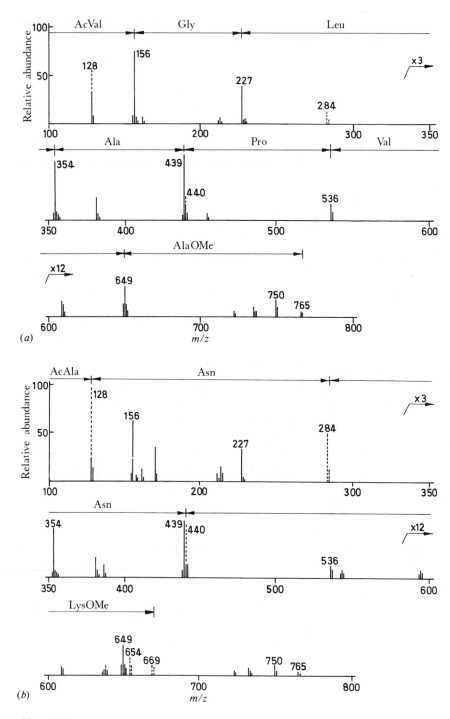

Figure 6-23

and looking for changes in peak abundance as one component becomes more or less intense relative to the other. A component which is more abundant in the scan at 200°C (Fig. 6-23a) has the sequence Val.Gly.Leu.Ala.Pro.Val.Ala; the second component, which gives rise to ions of increased abundance at 220°C (Fig. 6-23b), is a tetrapeptide of sequence Ala.Asn.Asn.Lys (see dotted lines in the figures). A third component, volatilizing at a temperature of 260°C, was shown to have the sequence Gln.Tyr.Tyr.Thr.Val.Phe, although the Phe residue was not the C terminal; additional residues beyond this point were present but not identified.

Further enzymic digests of the initially produced peptides which are too large to handle by mass spectrometry, coupled with the use of different enzymes for the initial digest (to create overlaps in sequence), provide a strategy for the sequencing of an enzyme by mass spectrometry.

The most useful achievements of mass spectrometry in peptide chemistry have been in (1) the (usually partial) sequencing of proteins,[24] (2) the identification of blocked N termini of peptides,[25] (3) the identification of unusual, or novel, amino acids,[26] and (4) the analysis of mixtures.[27] In this last role, mass spectrometry was used to sequence successfully a mixture of pentapeptides, which constitutes one of the brain's natural opiates. In tackling these kinds of problems, the usefulness of deuteriated reagents (CD_3I; d_6-Ac_2O; CD_3OH/HCl for esterification) in confirming conclusions drawn from the spectra of normal derivatives cannot be overemphasized.

This section has fully described most of the procedures for determining the sequence of amino acids in peptides by mass spectrometry. An important additional method is discussed in Sec. 10-1C, involving selective cleavage into small peptides, followed by derivatization and analysis by GC/MS.

b. Carbohydrates Much of the early work in this field has been reviewed by Kochetkov and Chizhov[28] and Heyns and coworkers.[29]

Carbohydrates are invariably derivatized prior to analysis by electron-impact mass spectrometry, and most work has been carried out on peracetates, pertrimethylsilyl ethers, or permethyl ethers. In the case of monosaccharides, structural information may, if necessary, be obtained by GC/MS following exhaustive trimethylsilylation. In oligosaccharides one would ideally wish to determine the number of sugars involved, their sequence, nature of linking (for example, 1 → 4, 1 → 6, etc.), and stereochemical details. The present situation is that molecular ions have been obtained for derivatized mono- to pentasaccharides and that prominent sequence ions frequently allow the ordering of monosaccharide units; using sugars with known modes of linking (for example, 1 → 4, 1 → 6, etc.), criteria for distinguishing between the various possibilities have been developed in some cases, but it is not yet clear how widely such criteria might be applied with confidence to sugars of unknown structure. In general, mass spectrometry has not proved sensitive to stereoisomeric differences between carbohydrates. However, a useful approach[30] is based on the fact that cyclic acetals and boronic esters of stereoisomers are frequently structural isomers and may give very different mass spectra. For example, the cases of D-galactose (**31**) and D-glucose (**32**) may be cited. The former has two pairs of *cis*-1,2-diol groupings in the α-pyranoid form and therefore forms mainly a pyranoid 1,2:3,4-di-O-isopropylidene derivative (**33**). D-Glucopyranose, on the other hand, has only one *cis*-diol group in the α form (**32**) (none in the β form), but in the otherwise less favorable furanoid α isomer **34** is capable of reacting with two molecules of acetone to form 1,2:5,6-di-O-isopropy-lidene-D-glucofuranose (**35**).

The partial mass spectra (above $m/z\,90$) of the isomers **35** and **33** are reproduced in Fig. 6-24a and Fig. 6-24b, respectively, and are clearly quite different. Note, for example, that only the 1,2:3,4-di-O-isopropylidene derivative loses acetic acid from the $M^{+}-15$ ion, and that the 1,2:5,6-di-O-isopropylidene derivative produces a more abundant ion at $m/z\,159$ due to the relatively facile formation of **t**.

Having illustrated how stereochemical information may be derived, it remains only to indicate the principles by which the numbers, sequence, and nature of linkage of monomer units may be obtained. Further to important work on di- and trisaccharides,[31] molecular ions of 1-phenylflavazole paracetates of di- to penta-saccharides have been obtained.[32] The mass spectrum of the 1-phenylflavazole peracetate (**36**) of maltopentose is reproduced in Fig. 6-25.

36 (Glc = peracetylated glucose ring)

The incorporation of the aromatic residue into the saccharide molecule represents an important application of earlier theoretical considerations. The

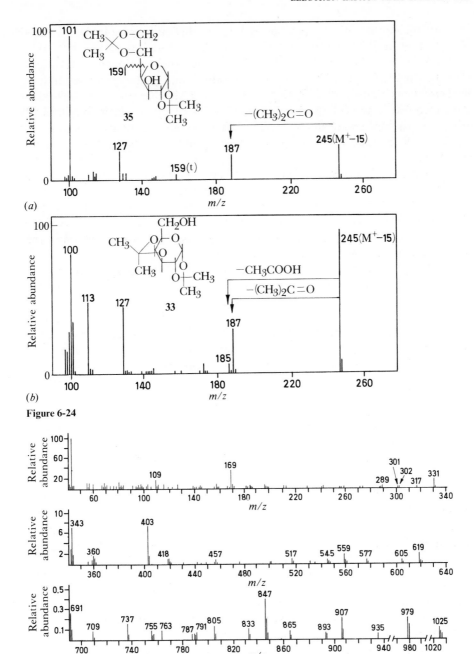

Figure 6-24

Figure 6-25

aromatic moiety has a low ionization energy and therefore some molecular ions are produced in which the total internal excitation energy is insufficient to cause ionization in the sugar part of the derivative. Hence, these molecular ions do not have available to them the low-energy decomposition pathways of sugar per-acetates which are ionized in the sugar portion, and M^+ ions survive long enough to be recorded at the collector. In addition, the derivatives, which are prepared using o-phenylenediamine and phenylhydrazine and purified using cellulose tlc[33] fluoresce; this can be an advantage when experiments are carried out on $\leqslant 1$ μmol of material.

The major sequence ion cleavages of the maltopentose derivative **36** are indicated in the structure. In the case where each sugar corresponds to a peracetylated glucose ring, as in the present case, these sequence ions differ in mass by 288 mass units. These ions, together with their daughters due to acetic acid (60 mass units) and/or ketene (42 mass units) loss, constitute a large fraction of the abundant ions above m/z 350 in the spectrum.

In the spectra of the tri- and tetrasaccharide 1-phenylflavole peracetates, it is further observed that when the first glycosidic linkage (numbering from the derivatized end of the molecule) is $1 \to 6$, the $[M^+ - 42]/[M^+]$ abundance ratio is high (for example, 9, 18, 30 in three representative samples) but low (for example, 0.4, 0.3, and 0.4) in compounds having $1 \to 4$ linkages. The nature of the second and subsequent glycosidic linkages is correlated with the ease with which acetic acid is lost by the fragment ions belonging to the sequence not retaining the flavazole (viz., m/z 619, 907, 1195 in **36**). In all cases, these ions lose acetic acid readily if and only if the lowest-numbered glycosidic linkage (adjacent to that cleaved in forming the ion in question) is $1 \to 4$. [Sequence ion $-$ AcOH]/[sequence ion] ratios of 1.0–2.5 are observed for $1 \to 4$ linkages, but the ratios for $1 \to 6$ linkages are 0.04–0.10.

The above method is of course restricted to oligosaccharides which can form 1-phenylflavazoles, but it illustrates well the general principles involved, and it should prove possible to modify the approach using other heteroaromatic groups[34] and other hydroxyl-protecting groups (for example, OMe replacing OAc).

c. Nucleotides, nucleosides, and bases The first study of the mass spectra of nucleosides was carried out by Biemann and McCloskey.[35] Molecular ions were obtained in numerous cases without derivatization of the parent nucleoside by vaporizing the samples (at the submicrogram level if necessary) about 1 cm from the ionizing electron beam. However, not all nucleosides afford molecular ions under these conditions, and therefore field ionization or field desorption is sometimes used (Sec. 7-3). Detailed studies have been undertaken of the mass spectra of pertrimethylsilyl derivatives of mononucleosides, mononucleotides, and bases,[36] and of pertrimethylsilyl derivatives of dinucleotides.[37] Base sequence information can be derived from the spectra of derivatized dinucleotides.[37,38]

However, since elegant and very sensitive "wet" chemical methods are available for the sequencing of nucleotides, it seems at the present that the main advantage of mass spectrometry in this area lies in the identification of unusual bases, frequently those isolated from transfer RNAs. For example, it was possible to isolate from an enzymic hydrolysis of yeast tRNA very small amounts of an unknown nucleoside containing impurities which stemmed from the isolation procedure.[39] In this case, it was advantageous to record a complete high-resolution spectrum; peaks from the nucleoside will contain relatively large numbers of oxygen and/or nitrogen atoms, which allow their differentiation from impurity peaks. Normally, mass spectra of nucleosides contain abundant ions due to B+H (and frequently also B+2H) where B represents the base of the nucleoside (see also Sec. 7-3 and Fig. 7-9). Indeed, the unknown afforded a pair of intense peaks of m/z 200 and 201, which surprisingly were shown from the exact mass measurements to contain sulfur ($C_7H_8N_2O_3S$ and $C_7H_9N_2O_3S$, respectively). Hence, it is possible to infer that the structure of the new nucleoside is represented by **37**; this formulation is supported by the presence of a $C_5H_9O_4$ ion corresponding to the pentoside unit.

37

38

The anticipated molecular ion species was not evident in the mass spectrum, but ions of the compositions $C_{11}H_{13}N_2O_6S$ and $C_{12}H_{14}N_2O_7$ (corresponding to $M^+ - CH_2OH$ and $M^+ - H_2S$) indirectly confirm the postulated molecular formula. The nitrogen content indicates a pyrimidine rather than a purine ring. In addition, ions with the lowest number of oxygen atoms correspond to $C_5H_5N_2OS$ and $C_5H_6N_2OS$, and these are compatible with a substituted thiouracil (for example, **38**), rather than with a substituted uracil. The minimum number of carbon atoms in the $C_nH_mN_2O_2S$ group is 6 and in the $C_nH_mN_2O_3S$ group is 7, and such data are most compatible with **38**. The placing of the side chain at carbon atom 5 (as opposed to carbon atom 6) is arbitrary on the basis of the mass spectrometry data, but is supported by ultraviolet spectra. The position of sulfur at carbon atom 2 (rather than carbon atom 4) is based upon H_2S loss upon electron impact being anticipated to be facile if it can proceed with formation of a carbon atom 2, carbon atom 5' ether bridge (**39**).

39

40

The complete structure **40** would of course be very tentative on the basis of mass spectrometric evidence alone, but is supported by ultraviolet spectra and chemical transformations (e.g., demonstration of the presence of a saponifiable group).

6-6 ORGANOMETALLIC COMPOUNDS

A. Isotope Patterns

Interpretation of the mass spectra of organometallic compounds is frequently complicated by the fact that many metals exist as a mixture of isotopes. This is especially true if more than one metal atom is present, but conversely a particular isotopic pattern may be useful in identifying one or more metal atoms. An example of the 70 eV spectrum of a multi-isotopic species, tetramethyltin, is shown in Fig. 6-26.[40] The characteristic isotopic pattern of tin, which has ten naturally occurring stable isotopes, is evident. Successive losses of methyl groups from the molecular ion are observed and some metal hydride species are found.

It can be envisaged from Fig. 6-26 that spectra will become very complicated if two or more multi-isotopic atoms are present, and will be complicated still further if two groups of ions of different composition are sufficiently close to overlap (e.g., the groups of ions having different compositions are almost overlapping in Fig. 6-26).

As a simple example of calculation of isotopic abundances in organometallic compounds, consider the molecular ion of $Re(CO)_5Cl$. Rhenium has two naturally occurring isotopes, ^{185}Re and ^{187}Re, of respective abundances 37.07 and 62.93

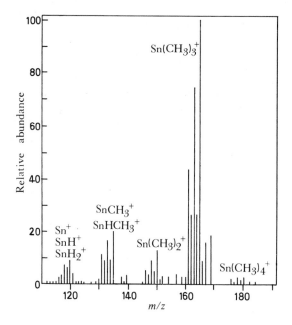

Figure 6-26

Nominal mass	Relative abundance	
398	18.4132	
399	2.8661	
400	35.4639	
401	30.6747	
402	70.4708	
403	49.9924	
404	93.0164	
405	68.2517	
406	100.0000	
407	51.7477	
408	78.3215	
409	20.0667	
410	32.7513	
411	4.8487	
412	4.7385	
413	0.6562	
414	0.0730	

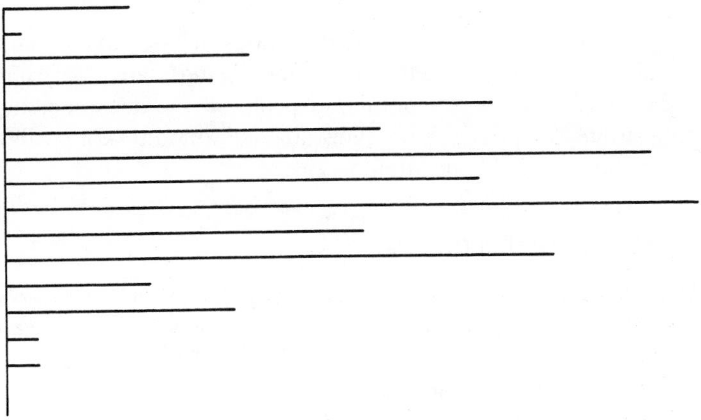

Figure 6-27

percent. The two naturally occurring isotopes of chlorine, ^{35}Cl and ^{37}Cl, have respective abundances of 75.53 and 24.47 percent. The ion $ReCl^{+}$ in the mass spectrum of $Re(CO)_5Cl$ would therefore be found at three different nominal m/z values of 220, 222, and 224. The abundance at each of the respective m/z values is obtained by multiplying together the abundances of the constituent atoms, that is, $^{185}Re^{35}Cl^{+}$ at m/z 220 will have an abundance (relative to the total abundance of $ReCl^{+}$) of 0.3707×0.7553 ($= 0.2800$). The abundances at m/z 222 and 224 are similarly obtained (m/z 222 has two components). The calculated abundances are: m/z 220, 28.0 percent; m/z 222, 56.6 percent; m/z 224, 15.4 percent. The calculation of peak distribution in the molecular ion $Re(CO)_5Cl^{+}$ is further complicated by the natural abundance (1.112 percent) of ^{13}C.

Computer programs may be written to facilitate the interpretation of the spectra of molecules containing more than one multi-isotopic atom. For example, the calculated peak distribution for the molecular ion of $\pi\text{-}C_5H_5(CO)_3MoGeEt_3$ is shown in Fig. 6-27.[41]

B. Thermal Decomposition

Another complication which may influence organometallic spectra is thermal decomposition in the mass spectrometer. It appears that many organometallic compounds are susceptible to such effects, and in these cases analytical and energetic data can be seriously affected by spurious features in the mass spectra. For example, at a heated inlet temperature of $150°C$, $Fe(CO)_5$ decomposes thermally to an extent of 20–30 percent to form mainly $Fe(CO)_4$ and CO.[42] As the temperature is increased, the abundances of $Fe(CO)_4^{+}$ and CO^{+} increase, whereas those of the other ions in the spectra decrease. Under conditions where decomposition occurs an appearance energy of 8.48 eV was measured for

$Fe(CO)_4{}^+$, whereas at low temperatures this was increased to 9.1 eV. It seems likely that the ionization energy of $Fe(CO)_4$ is the quantity measured in the former case.

As a general rule, thermal decomposition should be suspected in organometallic mass spectrometry when the ligand ions (for example, CO^+, NO^+, $C_5H_5{}^+$) give excessively dominant peaks. If the decomposition results from the deposition of catalytic metal films, a small amount of air will poison the catalyst and reduce the decomposition.

C. Fragmentation Processes

It is not proposed to attempt here any cataloging of the reactions of organometallic compounds, because of the enormous volume of relevant publications.[43,44] Instead, a few important and typical fragmentations have been selected to illustrate general principles.

The highly thermally stable metallocenes [$M(C_5H_5)_2$, where M is a metal] have low ionization energies and yield abundant molecular ions which fragment by successive loss of C_5H_5 units.[45] The stability of the molecular ions of metallocenes is reflected in the stability of the same ions in aqueous solution and in the solid phase. For example, cobaltocene [$Co(C_5H_5)_2$], having a 19-electron configuration, is oxidized with air in aqueous solution to form the cobalticinium ion, $Co(C_5H_5)_2{}^+$, which may be isolated as its $PF_6{}^-$ salt.

A large number of substituted metallocenes have been shown to undergo some unusual rearrangements, involving the organic side chain, in their mass spectra. For example, ejection of C_5H_4CO has been observed from the molecular ion of the substituted ferrocenes **41**, followed by further elimination of C_5H_5 (see **41 → 42 → 43**):[46]

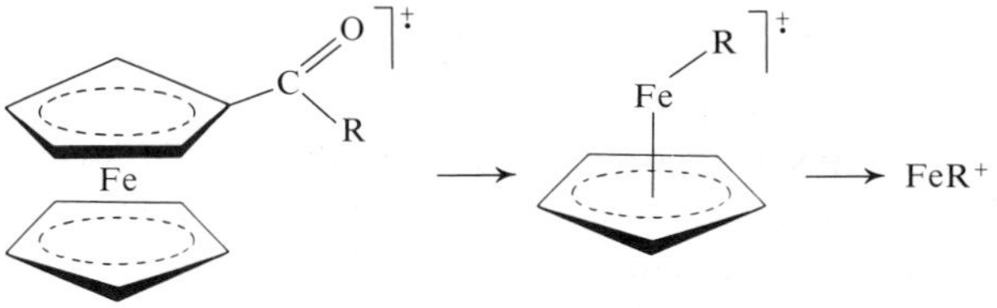

The importance of this rearrangement in the mass spectrum (normalized for ^{56}Fe and ^{12}C) of $C_5H_5-Fe-C_5H_4CO_2H$ is evident in Fig. 6-28, where the base peak corresponds to ion **42**.

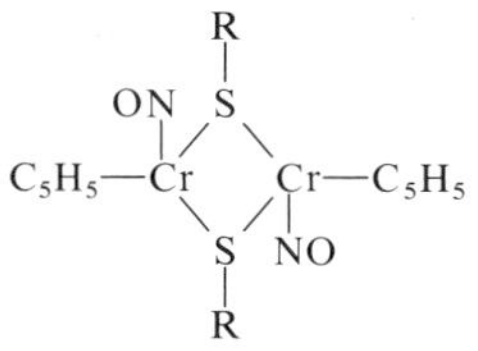

44

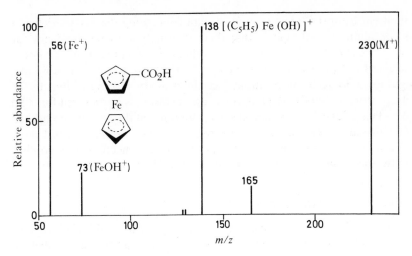

Figure 6-28

Migrations of ligands are also relatively common in the breakdown of organometallic ions. For example, the spectra of the binuclear metal nitrosyls with sulfur bridges (**44**, R = Me, Ph) exhibit an ion at m/z 182 corresponding to the composition $Cr(C_5H_5)_2^+$ [47] Migration of CO is known in the breakdown of binuclear carbonyl molecular ions, and caution should be exercised in structural interpretation that more metal—CO bonds than are actually present are not attributed to one of the metal atoms.

The mass spectrum of iron pentacarbonyl (Fig. 6-29) illustrates the stepwise loss of a ligand from a molecular ion.[48] Ions are observed for all the species $Fe(CO)_n^+$ ($n = 0-5$) and metastable peaks are observed for all the relevant transitions.[49] Energetic data for these reactions are available.[50]

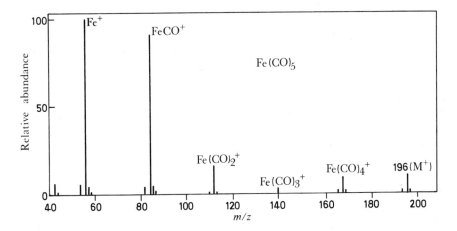

Figure 6-29

It is apparent from the large amount of data concerning the mass spectra of complexes containing mixed ligands that there is usually a greater tendency for some ligands to be eliminated in preference to others. For example, from studies of nitrosyl carbonyl complexes, it is evident that CO groups are more readily lost than NO groups in ionic decompositions [see the primary decompositions in the mass spectrum (Fig. 6-30) of $Fe(CO)_2(NO)_2$; the spectrum is normalized for ^{56}Fe and $^{12}C].[51]$ This is due to a greater electron-donating ability of NO compared with the CO ligand. When the strong electron-donating substituent $P(OEt)_3$ is introduced into nitrosyl carbonyl complexes, a stronger metal–ligand bond is formed and the metal preferentially retains the $P(OEt)_3$ ligand.

D. Detection of Polymeric Species

Mass spectrometry is a useful technique for detecting the presence of dimeric or polymeric species in organometallic compounds. For example, a number of instances are known of association (through bridging bonds) in solution of electron-deficient organometallics, e.g., the aluminum alkyls. The mass spectra of Me_3Al, Me_2AlH, Et_2AlH, and Et_2AlOEt have all revealed evidence for dimers and trimers, particularly at low source temperatures. For example, the highest mass ion in the spectrum of trimethylaluminum corresponds to $Me_5Al_2{}^+$ and carries 3 percent of the total ion current at a source temperature of $50°C.[52]$ This ion is presumably generated by elimination of a methyl radical from the ionized dimeric species **45**.

45

6-7 SKELETAL REARRANGEMENT

It was shown in the previous section that group migrations are quite common in the mass spectra of organometallic compounds. Such rearrangements (generally known as *skeletal rearrangements*) are less common in the spectra of ordinary organic compounds, but do occur with moderate frequency. A large number of examples have been documented in two review articles.[53,54]

Frequently group migration seems to be initiated by the generation of an electron-deficient site upon ionization or fragmentation. An isomerized molecular ion, or fragment ion, is then generated. For example:

$$\overset{+}{A}-B-C \longrightarrow C-A-\overset{+}{B} \longrightarrow C-\overset{+}{A} \quad +B$$

To exemplify this kind of behavior, an example is taken from the numerous cases of migrations to carbonium ion centers. In the mass spectrum (Fig. 6-31) of methyl α-bromophenylacetate (**46**), the $M^+ - Br$ ion undergoes loss of CO in a one-step

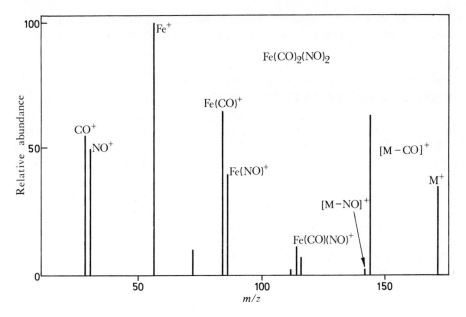

Figure 6-30

process with associated methoxyl migration, presumably to the carbonium ion center.[55] In the following representation of the rearrangement, the structures of the ions are assumed and not proven:

$$C_6H_5CH(Br)CO_2Me \xrightarrow[-Br]{-e} C_6H_5\overset{+}{C}H-C\overset{OMe}{\underset{O}{\diagdown}} \xrightarrow[*]{-CO} C_6H_5\overset{+}{C}HOMe$$

46 $m/z\ 121$

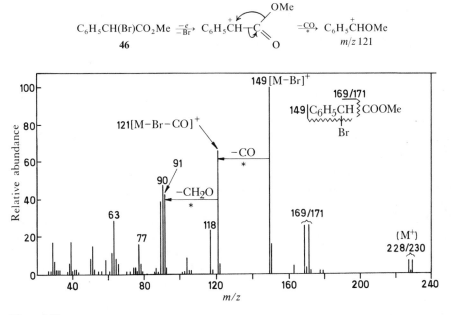

Figure 6-31

Such rearrangements do, of course, generate units which are not present in the original molecule, and may on occasions lead to incorrect structural conclusions being made from a mass spectrum.

6-8 TEMPERATURE AND ION ABUNDANCES

Mass spectra can, of course, be greatly simplified on reduction of the electron-beam energy from 70 eV to a nominal 12–15 eV. At low beam energies, only the primary reactions of low activation energy prevail, and consecutive decompositions are minimized (Fig. 3-12). However, it remained for Spiteller and coworkers to show that in the case of large molecules the effect of source temperature upon relative ion abundances is also considerable.[56,57] The simplest mass spectra are obtained by using a "cold" ion source ($\sim 70^{\circ}$C) at low voltage (say, 12 eV). Particularly noteworthy is the fact that a temperature decrease of 250°C may, in certain classes of compounds, lead to an increase in the intensity of the parent ion peaks by a factor of 200–300 compared to those recorded at the higher temperature. In general, the smaller the amount of energy required for the cleavage of the parent ion, the greater is the influence of temperature upon the M^+ abundance. For example, the abundance of the molecular ion peak in the mass spectrum of the saturated hydrocarbon triacontane ($C_{30}H_{62}$) is of the order of 1 percent of the m/z 57 ($C_4H_9^+$) abundance at 70 eV and 340°C, but is 86 percent of the m/z 57 abundance at 70 eV and 70°C. Although such large variations are somewhat unusual, and might seem surprising at first glance, the effects are more readily understandable if a calculation of the thermal energy of $C_{30}H_{62}$ at 340°C is made; one obtains a value of about 3 eV ($1 \text{ eV} = 23.06 \text{ kcal mol}^{-1} = 96.49 \text{ kJ mol}^{-1}$). In a large molecule, the thermal energy is significant relative to the average internal energy gained upon electron impact with 70 eV electrons (see Fig. 2-9).

6-9 "IN-BEAM" MASS SPECTROMETRY

There are many indications that samples may be more efficiently volatilized by methods other than direct heating. One such indication is the remarkably increased molecular ion abundances obtained from relatively involatile samples when the probe tip is "fringed" by the electron beam. In this technique, the sample is deposited on the outside of the direct insertion probe tip, and the probe tip then advanced until it just "fringes" the electron beam.[58] The tip should not extend into the main part of the electron beam since the electron-beam current would then fall; in some instruments, an automatic increase in filament temperature would follow (to attempt to restore the electron-beam current to its normal value) with the possible consequence that the filament would "burn out". It appears that when the sample is bombarded with electrons, volatilization at the expense of thermal degradation is improved.

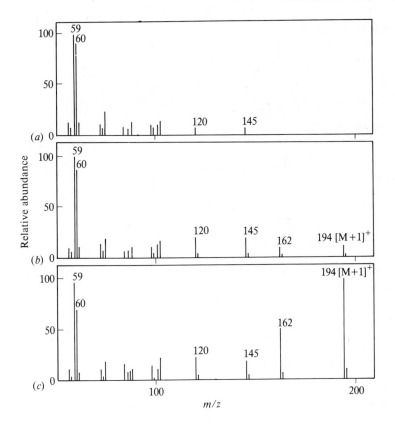

Figure 6-32

Spectra obtained in the above manner may be dominated by either $M^{+[58]}$ or $[M+1]^{+[59]}$ ions. The change in the spectrum of methyl-3-deoxy-3-amino-α-D-glucopyranoside (**47**) with variation of the distance between the sample and the electron beam is given in Fig. 6-32 for (*a*) 15 mm (conventional electron impact), (*b*) 3 mm, and (*c*) 0 mm.

47

Related useful effects have been obtained by inserting the sample into a chemical ionization plasma,[60] or loading the sample on a field desorption emitter into a chemical ionization plasma,[61] or simply by using the field desorption emitter as a solid sample probe to obtain improved electron-impact spectra.[62]

REFERENCES

1. K. Biemann: "Mass Spectrometry," McGraw-Hill, New York, 1962.
2. H. Budzikiewicz, C. Djerassi, and D. H. Williams: "Structure Elucidation of Natural Products by Mass Spectrometry," vols. 1 and 2, Holden-Day, San Francisco, 1964.
3. F. W. McLafferty: "Interpretation of Mass Spectra," W. A. Benjamin, New York, 1966.
4. H. Budzikiewicz, C. Djerassi, and D. H. Williams: "Mass Spectrometry of Organic Compounds," Holden-Day, San Francisco, 1967.
5. J. H. Beynon, R. A. Saunders, and A. E. Williams: "The Mass Spectra of Organic Molecules," Elsevier, Amsterdam, 1968.
6. (a) J. H. Beynon: "Mass Spectrometry and its Applications to Organic Chemistry," pp. 262–263, Elsevier, Amsterdam, 1960; (b) G. Spiteller: "Massenspectrometrische Structuranalyse Organische Verbindungen," pp. 88, 97–98, Verlag Chemie, Weinheim, 1966.
7. M. Kraft and G. Spiteller: *Org. Mass Spec.*, vol. 2, p. 541, 1969.
8. F. W. McLafferty: *Anal. Chem.*, vol. 34, p. 2, 1962.
9. G. Remberg, E. Remberg, M. Spiteller-Friedmann, and G. Spiteller: *Org. Mass Spec.*, vol. 1, p. 87, 1968; G. Remberg and G. Spiteller: *Chem. Ber.*, vol. 103, p. 3640, 1970.
10. C. Djerassi: *Pure Appl. Chem.*, vol. 6, p. 575, 1963.
11. K. Biemann and G. Spiteller: *Tetrahedron Letters*, p. 299, 1961; *J. Am. Chem. Soc.*, vol. 84, p. 4578, 1962.
12. B. Gilbert, J. A. Brissolese, J. M. Wilson, H. Budzikiewicz, L. J. Durham, and C. Djerassi: *Chem. and Ind.*, p. 1949, 1962.
13. H. Audier, A. Diara, M. de J. Durazo, M. Fetizon, P. Foy, and W. Vetter: *Bull. Soc. Chim. Fr.*, p. 2827, 1963; Z. Pelah, D. H. Williams, H. Budzikiewicz, and C. Djerassi: *J. Am. Chem. Soc.*, vol. 86, p. 3722, 1964.
14. G. P. Arsenault, K. Biemann, A. W. Barksdale, and T. C. McMorris: *J. Am. Chem. Soc.*, vol. 90, p. 5635, 1968.
15. J. A. Edwards, J. S. Mills, J. Sundeen, and J. H. Fried: *J. Am. Chem. Soc.*, vol. 91, p. 1248, 1969.
16. M. Barber, P. Jollès, E. Vilkas, and E. Lederer: *Biochem. Biophys. Res. Comm.*, vol. 18, p. 469, 1965.
17. B. C. Das, S. D. Gero, and E. Lederer: *Biochem. Biophys. Res. Comm.*, vol. 29, p. 211, 1967.
18. D. W. Thomas, B. C. Das, S. D. Gero, and E. Lederer: *Biochem. Biophys. Res. Comm.*, vol. 33, p. 519, 1968.
19. D. W. Thomas: *FEBS Letters*, vol. 5, p. 53, 1969.
20. H. R. Morris: *FEBS Letters*, vol. 22, p. 257, 1972.
21. H. R. Morris, D. H. Williams, and R. Ambler: *Biochem. J.*, vol. 125, p. 189, 1971.
22. J. V. Stone, W. Mordue, K. E. Batley, and H. R. Morris: *Nature*, vol. 263, p. 207, 1976.
23. F. W. McLafferty, R. Venkataraghavan, and P. Irving: *Biochem. Biophys. Res. Comm.*, vol. 39, p. 274, 1970.
24. H. R. Morris, D. H. Williams, G. G. Midwinter, and B. S. Hartley: *Biochem. J.*, vol. 141, p. 701, 1974.
25. A. D. Auffret, D. H. Williams, and D. R. Thatcher: *FEBS Letters*, vol. 90, p. 324, 1978.
26. H. R. Morris, A. Dell, T. E. Petersen, L. Sottrup-Jensen, and S. Magnusson: *Biochem. J.*, vol. 153, p. 663, 1976.
27. J. Hughes, T. W. Smith, H. W. Kosterlitz, L. A. Fothergill, B. A. Morgan, and H. R. Morris: *Nature*, vol. 258, p. 577, 1975.
28. N. K. Kochetkov and O. S. Chizhov: *Adv. Carbohydrate Chem.*, vol. 21, p. 39, 1966.
29. K. Heyns, H. F. Grützmacher, H. Scharmann, and D. Müller: *Fortschr. Chem. Forsch.*, vol. 5, p. 448, 1966.
30. D. C. De Jongh and K. Biemann: *J. Am. Chem. Soc.*, vol. 86, p. 67, 1964.
31. N. K. Kochetkov, O. S. Chizhov, and N. V. Molodtsov: *Tetrahedron Letters*, vol. 24, p. 5587, 1968.
32. G. S. Johnson, W. S. Ruliffson, and R. G. Cooks: *Chem. Comm.*, p. 587, 1970.
33. P. Nordin: *Methods Carbohydrate Chem.*, vol. 2, p. 136, 1963.
34. N. K. Kochetkov, O. S. Chizhov, N. N. Malysheva, and A. I. Shiyonok: *Org. Mass Spec.*, vol. 5, p. 481, 1971.

35. K. Biemann and J. McCloskey: *J. Am. Chem. Soc.*, vol. 84, p. 2005, 1962; see also S. M. Hecht, A. S. Gupta, and N. J. Leonard: *Analyt. Biochem.*, vol. 30, p. 249, 1969.
36. A. M. Lawson, R. N. Stillwell, M. M. Tacker, K. Tsuboyama, and J. A. McCloskey: *J. Am. Chem. Soc.*, vol. 93, p. 1014, 1971.
37. D. F. Hunt, C. E. Hignite, and K. Biemann: *Biochem. Biophys. Res. Comm.*, vol. 33, p. 378, 1968.
38. J. J. Dolhun and J. L. Wiebers: *J. Am. Chem. Soc.*, vol. 91, p. 7755, 1969.
39. L. Baczynskyj, K. Biemann, and R. H. Hall: *Science*, vol. 159, p. 1481, 1968.
40. R. W. Kiser: In M. Tsutsui (ed.), "Characterisation of Organometallic Compounds," pt. I, p. 147, Wiley-Interscience, New York, 1969.
41. A. Carrick and F. Glockling: *J. Chem. Soc.*, vol. A, p. 40, 1967.
42. S. Pignataro and F. P. Lossing: *J. Organometallic Chem.*, vol. 11, p. 571, 1968.
43. M. R. Litzow and T. R. Spalding: "Mass Spectrometry of Inorganic and Organo-metallic Compounds," Elsevier, Amsterdam, 1973.
44. *Specialist Periodical Reports: Mass Spectrometry*, vols. 1–5, The Chemical Society, London, 1971, 1973, 1975, 1977, 1979.
45. L. Friedman, A. P. Irsa, and G. Wilkinson: *J. Am. Chem. Soc.*, vol. 77, p. 3689, 1955.
46. A. Mandelbaum and M. Cais: *Tetrahedron Letters*, vol. 51, p. 3847, 1964.
47. F. I. Preston and R. I. Reed: *Chem. Comm.*, p. 51, 1966.
48. R. E. Winters and R. W. Kiser: *Inorg. Chem.*, vol. 3, p. 699, 1964.
49. R. E. Winters and J. H. Collins: *J. Phys. Chem.*, vol. 70, p. 2057, 1966.
50. J. L. Franklin et al.: "Ionization Potentials, Appearance Potentials and Heats of Formation of Gaseous Positive Ions," NSRDS–NBS 26, U.S. Govt. Printing Office, 1969.
51. A. Foffani, S. Pignataro, G. Distefano, and G. Innorta: *J. Organometallic Chem.*, vol. 7, p. 473, 1967.
52. D. B. Chambers, G. E. Coates, F. Glockling, and M. Weston: *J. Chem. Soc.*, vol. A, p. 1712, 1969.
53. P. Brown and C. Djerassi: *Angew. Chem. Int. Edn.*, vol. 6, p. 477, 1967.
54. R. G. Cooks: *Org. Mass Spec.*, vol. 2, p. 481, 1969.
55. R. G. Cooks, J. Ronayne, and D. H. Williams: *J. Chem. Soc.*, vol. C, p. 2601, 1967.
56. M. Spiteller-Friedmann, S. Eggers, and G. Spiteller: *Monatsh.*, vol. 95, p. 1740, 1964.
57. M. Spiteller-Friedmann and G. Spiteller: *Chem. Ber.*, vol. 100, p. 79, 1967.
58. A. Dell, D. H. Williams, H. R. Morris, G. A. Smith, J. Feeney, and G. C. K. Roberts: *J. Am. Chem. Soc.*, vol. 97, p. 2497, 1975.
59. M. Ohashi, S. Yamada, H. Kudo, and N. Nakayama: *Biomed. Mass Spec.*, vol. 5, p. 578, 1978.
60. M. A. Baldwin and F. W. McLafferty: *Org. Mass. Spec.*, vol. 1, p. 1353, 1973.
61. D. F. Hunt, J. Shabanowitz, F. K. Botz, and D. A. Brent: *Anal. Chem.*, vol. 49, p. 1160, 1977.
62. B. Soltmann, C. C. Sweeley, and J. F. Holland: *Anal. Chem.*, vol. 49, p. 1164, 1977.

ALTERNATIVE METHODS OF IONIZATION.
APPLICATIONS

The wide range of applications of electron-impact (EI) mass spectrometry has been amply illustrated by the examples quoted in Chap. 6. However, even this highly successful and sensitive technique has its limitations, the main ones being concerned with (1) sample involatility and (2) molecular ion instability. The first problem arises with polar compounds, especially at high molecular weight. Even at the low pressure employed in the mass spectrometer it is impossible to volatilize some compounds from the heated probe and meaningful EI spectra are unobtainable. The second problem arises when the activation energy for unimolecular decomposition of the molecular ion is (almost) zero, yielding a spectrum containing fragment ions with a molecular ion of zero or low intensity. In such cases the molecular weight of a sample is difficult to identify.

However, as we have seen in Chaps. 1 and 2, various alternative ionization techniques have become available to provide complementary structural information to EI mass spectrometry. In this chapter some of the applications of these techniques are illustrated.

7-1 POSITIVE-ION CHEMICAL IONIZATION (PICI)

The type of information obtained from a PICI spectrum is illustrated in Fig. 7-1.[1] This figure shows one EI and two CI (methane and isobutane) spectra of the common plasticizer di-isooctylphthalate. The first point to note is the lack of a molecular ion in the EI spectrum (Fig. 7-1a). The spectrum is dominated by the m/z 149 ion which is a common feature of the spectra of the dialkylphthalate family of plasticizers (found, for example, in tlc plates). In fact it is not always easy to distinguish members of this compound class by EI mass spectrometry.

The situation in the CI spectrum, however, is quite different. There is an abundant MH^+ ion at m/z 391 whose fragmentation is postulated below:

The following points are evident from the above spectra:

1. The CI spectra enable the molecular weight to be determined and therefore (in this case) allow the different compounds of the same chemical class to be distinguished from one another.
2. There is less fragmentation in the isobutane CI spectrum (Fig. 7-1*c*) compared with the methane spectrum (Fig. 7-1*b*) because the $C_4H_9^+$ ion is a milder Brönsted acid than CH_5^+ (as discussed in Sec. 2-2A).

It was mentioned in Sec. 2-2 that the NH_4^+ ion, produced in the ammonia CI plasma at pressures around 1 torr, is a useful mild ionizing reagent, acting either via proton transfer or via solvation of the neutral sample molecule. Ammonia CI is now widely used in the natural product field, particularly for molecules containing sugar moieties.

The ammonia CI spectrum of the disaccharide sucrose is reproduced in Fig. 7-2. Meaningful EI spectra of underivatized sugars are unobtainable, whereas the sucrose CI spectrum shows an abundant $[M+NH_4]^+$ ion at m/z 360 which fragments characteristically by H_2O loss and to yield ions corresponding to the monosaccharide units.

The application of ammonia CI to the identification and analysis of sugars has been extended to glycosides and (with limited success) to higher underivatized oligosaccharides. However, with oligosaccharides containing three or more units it is preferable to prepare a suitable derivative because of the lack of volatility.

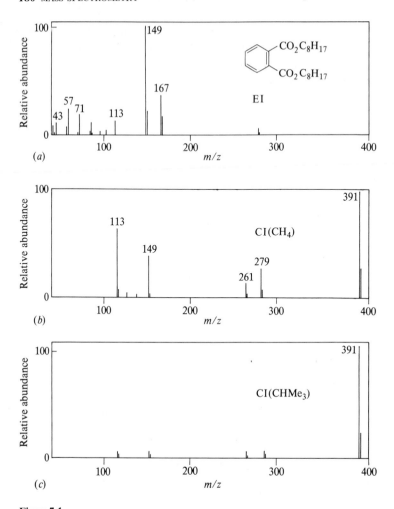

Figure 7-1

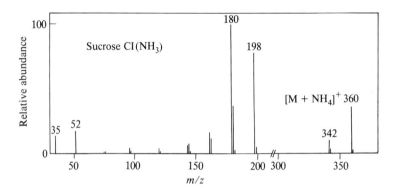

Figure 7-2

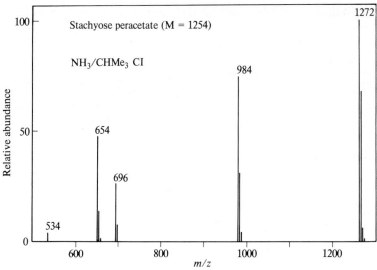

Figure 7-3

Furthermore, as CI techniques have become more sophisticated the use of reagent gas *mixtures* has emerged. It has thus become possible, where time and facilities permit, to investigate the optimum reagent gas mixture and pressure for good CI spectra of a particular compound class. In fact, it is not surprising from the organic chemist's experience of solution reactions that this possibility for the improvement of CI spectra exists by modifying the bimolecular ion chemistry in the ion source.

An example of the use of a reagent gas mixture is shown in Fig. 7-3. This figure shows the partial mass spectrum of a peracetylated tetrasaccharide, stachyose, obtained with a 2:1 mixture of ammonia and isobutane at 0.3 torr.[2] The $[M + NH_4]^+$ ion at m/z 1272 is the base peak and ions are found which correspond to the cleavage of glycosyl linkages.

When the identification of saccharides of unknown structure is necessary, then deuteriated reagents may be employed conveniently in the preparation of a derivative. Thus, an antibiotic suspected to contain a tetrasaccharide was subjected to mild hydrolysis.[3] The mixture of products so produced was permethylated using (1) $^-CH_2SOCH_3$ followed by treatment with CH_3I and (2) $^-CH_2SOCH_3$ followed by treatment with CD_3I. The crude chloroform extracts from the reactions were subjected to ammonia CI and the spectra reproduced in Fig. 7-4 were obtained. Figure 7-4a reproduces spectra obtained at three different source temperatures. In early (low-temperature) scans m/z 472 is observed; in later (medium-temperature) scans m/z 614 and 646 are produced; and finally, at the highest source temperatures at which ions are obtained from the sample, m/z 774 and 806 appear. When CD_3I replaced CH_3I in the derivatization process, the corresponding peaks appearing in Fig. 7-4b were obtained. Thus, the ions corresponding to m/z 472, 614, 646, 774, and 806 in Fig. 7-4a contain, respectively, 8, 9, 10, 11, and 12 methyl groups derived from methyl iodide. Since permethylated sugars are known to produce abundant "molecular ions" due to solvation of $\overset{+}{N}H_4$, it may be concluded that m/z 472, 646, and 806 correspond, respectively, to di-, tri-, and tetrasaccharide "molecular ions" containing 8, 10, and 12 methylatable sites. The m/z 614 and 774 ions then arise by loss of MeOH from the "molecular ions" of the tri- and tetrasaccharides. The data are satisfied if the three products of hydrolysis are a hexosyl-hexose, a deoxyhexosyl-hexosyl-hexose, and a pentosyl-deoxyhexosyl-hexosyl-hexose. Details of the manner of linkage of the sugars and stereochemistry are not revealed in these experiments. However, these experiments pointed to the identity of the antibiotic (ristocetin A) with another antibiotic (ristomycin A), known to contain the tetrasaccharide **1**. The identity of the two antibiotics was then confirmed by other methods.[3]

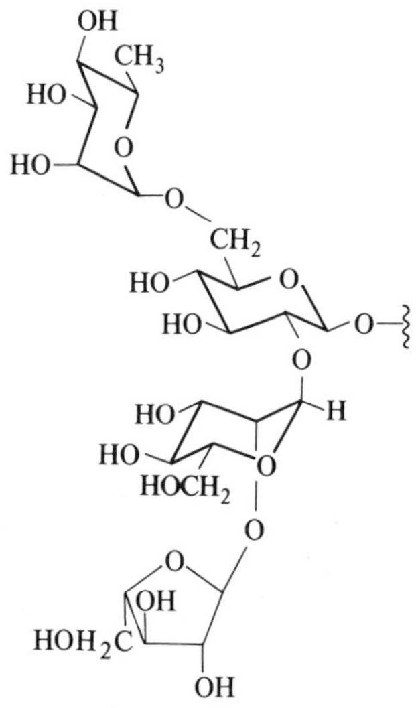

1

Since CI frequently produces stable ions of low energy it might be expected that this technique would stand a better chance of distinguishing between isomers (even stereoisomers) where EI mass spectrometry fails to achieve this. There are now a considerable number of examples to support this postulate.

Methane PICI mass spectrometry readily distinguishes between the isomers benzyl chloride and p-chlorotoluene, as shown in Fig. 7-5. The EI mass spectra of these two compounds are quite similar, containing a base peak at m/z 91 (the benzyl and/or tropylium ion) and differing only to a minor extent in the molecular ion intensity. This phenomenon may be explained in terms of reversible isomerization between the two molecular ion structures prior to decomposition.

The spectra shown in Fig. 7-5, however, are readily distinguishable from each other.[4] The difference between the spectra is due to the relatively large activation energy required for unimolecular fragmentation of any protonated form of p-chlorotoluene, whereas benzyl chloride protonated on chlorine can decompose readily to the benzyl cation.

$$C_6H_5CH_2\!-\!\overset{+}{C}lH \rightarrow C_6H_5CH_2^+ + HCl$$

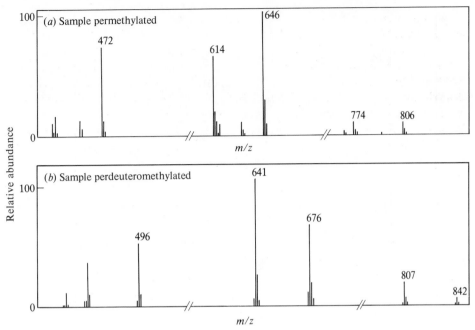

Figure 7-4

Incidentally, the spectrum shown in Fig. 7-4*b* contains an $[M+C_2H_5]^+$ ion, which is a common feature of methane CI spectra.

CI mass spectrometry has also been used to distinguish between stereo-isomers, particularly in alicyclic compounds containing two or more alcohol functions. Steric or stereoelectronic arguments readily accommodate the results. The possibility also arises that optical isomers might undergo different reactions in a CI source with a suitably chosen chiral ion.

The sensitivity of CI fragmentations to stereoelectronic control is illustrated by the contrasting CI spectra of the cis and trans stereoisomers **2** and **3**.[5]

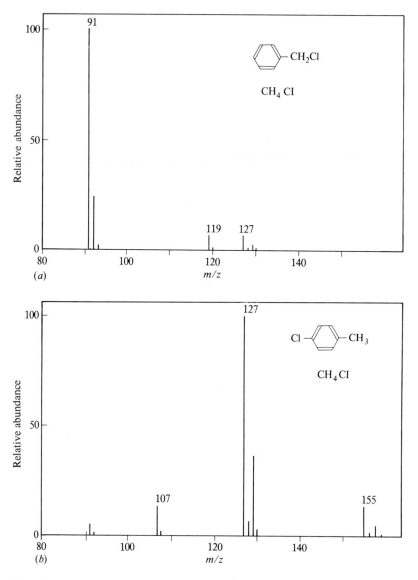

Figure 7-5

The former affords an abundant $[M+H]^+$ ion, whereas **3** does not show a corresponding ion in its spectrum, but rather an extremely abundant fragment ion due to loss of MeOH. There appears to be little doubt that this fragment ion arises because the $[M+H]^+$ ion **4** is ideally constituted to undergo a gas-phase intramolecular "S_N2" reaction (**4** → **5**). As seen earlier, gas-phase S_N2 reactions are noteworthy for their small or zero activation energies (Sec. 5-4B). The understanding of the gross differences between the mass spectra of **2** and **3** is sufficiently reliable that analogous reasoning can be used for the assignment of stereochemistry in similar cases. It is emphasized that since CI fragmentation usually derives from protonated molecules, a good knowledge of the requirements for facile S_N1 and S_N2 reactions is useful in the interpretation of CI spectra.

Although electron-impact mass spectrometry has proved so useful in the sequence determination of small peptides (Sec. 6-5D), CI spectra can give important complementary information.[6] There is an increased probability of obtaining the molecular weight of N-acylated-permethylated peptides (from the $[M+H]^+$ ion). Acyl ion fragments **a** are formed (and give sequence information in exactly the same way as in EI spectra). Additionally, ammonium ions **b** are produced; these ions allow the peptide to be sequenced from the C-terminal end. Although ions corresponding to **b** are rarely abundant for fragments containing only one or two amino acids, they are often abundant when **b** contains more amino acids. The availability of sequence information from both N- and C-terminal ends of the peptide allows an independent check of the structural conclusions based on only one set of data.

For example, in the hydrogen CI spectrum of an N-acetylated-permethylated octapeptide (H_3O^+ is the effective reagent ion, due to the presence of traces of H_2O in the source), ions at m/z 158, 271, 400, 497, 624, 695, and 780 are of type **a** and correspond to the sequence Ser.Val.Thr.Pro.Leu.Gly.Ala (Table 6-9). Abundant ions also occur at m/z 574, 703, and 816, indicating X^+, $ThrX^+$, and $Val.ThrX^+$. Since X^+ must correspond to $^+H_2Pro.Leu.Gly.AlaY—OMe$, where Y is the as yet undetermined part of the molecule, it is concluded that the derivatized mass of Y is 161 daltons. From Table 6-9, this is therefore identified as phenylalanine (Phe). Thus, even in the absence of a molecular ion in this spectrum, the complete sequence can be determined. The relationship of the sequence ions to the structure is given in **6**; note that the sum of the masses of the sequence ions of type **a** and **b** associated with cleavage of any one peptide bond is 2 daltons greater than the molecular weight.

6

Chemical ionization therefore has become a routine and valuable alternative ionization mode. Its main advantages are its selectivity and sensitivity, i.e., CI spectra frequently contain abundant ions in the molecular ion region and the total ion current compares favorably with that obtained from EI mass spectrometry. CI is therefore suitable for multiple ion detection and quantitative measurements. An advantageous feature of modern CI sources is the ability to switch rapidly (in several seconds) from CI to EI, and vice versa, thus enabling both CI and EI spectra to be obtained from one GC peak. The main disadvantage of CI mass spectrometry is the lack of structurally informative fragment ions on occasions. It should also be borne in mind that CI sources sometimes have short lifetimes, particularly where hydrocarbon reagent ions are employed.

7-2 NEGATIVE-ION CHEMICAL IONIZATION (NICI)

It has already been seen (Chaps. 1 and 2) that negative ions can be produced, analyzed, and recorded with relatively high efficiency. It was emphasized (Sec. 2-2C) that if a sample can be induced to form a molecular anion on nearly every encounter with a thermal electron, then high-pressure electron-capture should be two orders of magnitude more sensitive than EI or CI. Indeed, single-ion electron capture should allow detection of suitable samples in the low femtogram $(10^{-15} g)$ region.[7]

To achieve the above-mentioned sensitivity, the sample must naturally possess, or acquire through the preparation of a suitable derivative, a positive electron affinity. Positive electron affinities are observed for many halogenated compounds, quinones, and nitro compounds. Trace quantities of impurities with high electron affinities (e.g., molecules containing halogens) can seriously deplete the population of electrons in the ion source available for sample ionization. A sharp drop in sample sensitivity results,[7] and for this reason halocarbons are undesirable solvents for use in electron-capture studies.

Since spectra produced by electron capture in a high-pressure source often show relatively little fragmentation, it is often desirable to obtain a mass-marked

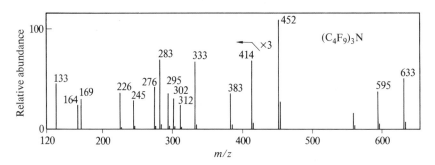

Figure 7-6

spectrum. In double-beam instruments (where separate ion beams are generated from different samples in two adjacent sources, analyzed in a single analyzer, and then separately recorded on adjacent detectors; see also Sec. 7-3), perfluorokerosene (PFK) can be used as a mass marker in the second beam. The PFK (1 μl) and air (4 ml) are introduced into a heated reservoir probe. This mixture is then bled into the second source. The resultant intermediate source pressure reduces the energy of some electrons to near thermal levels (via collisions) and allows collisional stabilization of negative ions formed from PFK by capture of the near thermal electrons.[8] The resulting spectrum of PFK is very useful for mass marking. Perfluorotributylamine may be similarly employed; its high-pressure negative-ion mass spectrum is reproduced in Fig. 7-6.

Since amino compounds readily form 2,4-dinitrophenyl derivatives upon reaction with 2,4-dinitrofluorobenzene, these derivatives (which exhibit the desired electron affinity) may be used to aid the identification of amino compounds by the above methods. The relevant spectrum of the 2,4-dinitrophenyl derivative of phenylalanylvaline methyl ester is reproduced in Fig. 7-7. It shows a large molecular anion peak and a smaller $[M - NO]^-$ ion. Contamination of the source by traces of iodine is indicated by m/z 127 (I^-) and 277 (probably NaI_2^-).

The utility of perfluorophenyl derivatives of organic compounds for such work has been reported.[7] For example, an amine may be condensed with pentafluorobenzaldehyde to give a pentafluorobenzylimine [Eq. (7-1)], an alcohol or phenol reacted with pentafluorobenzoyl chloride [Eq. (7-2)], or amines similarly acylated [Eq. (7-3)]:

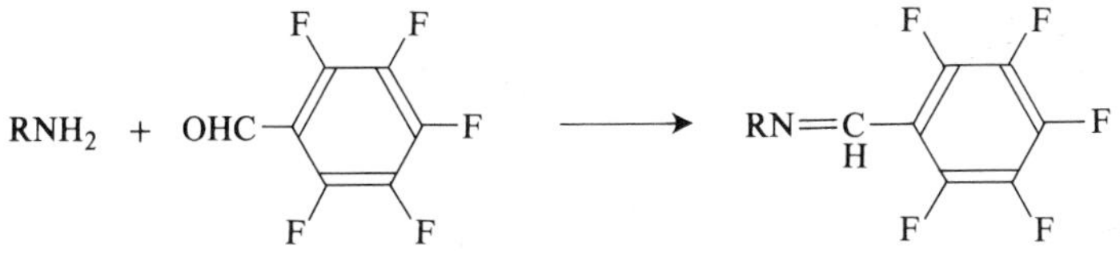

$$RNH_2 + OHC-C_6F_5 \longrightarrow RN{=}\overset{}{\underset{H}{C}}-C_6F_5 \qquad (7\text{-}1)$$

$$ROH + Cl\overset{O}{\overset{\|}{C}}-C_6F_5 \longrightarrow RO\overset{O}{\overset{\|}{C}}-C_6F_5 \qquad (7\text{-}2)$$

$$RNH_2 + Cl\overset{O}{\overset{\|}{C}}-C_6F_5 \longrightarrow R\underset{H}{N}\overset{O}{\overset{\|}{C}}-C_6F_5 \qquad (7\text{-}3)$$

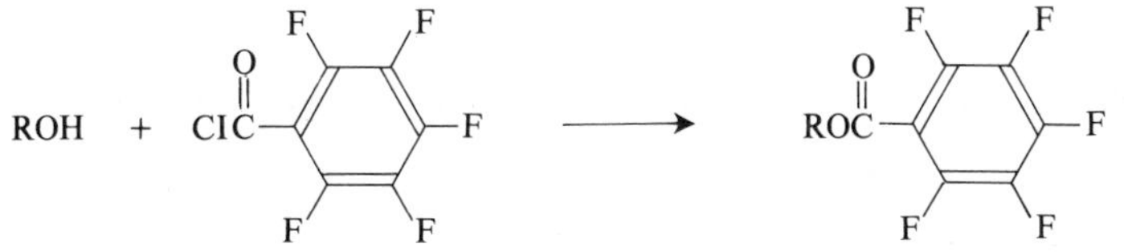

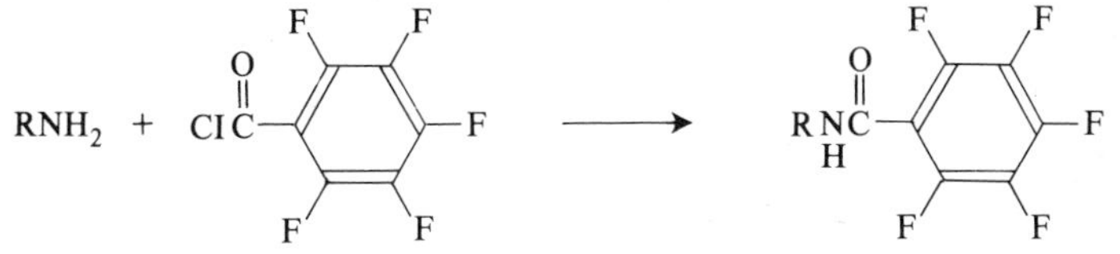

Using electron-capture derivatives of this kind, some organic substrates could be detected by single-ion monitoring at the 2.5×10^{-14} g level with a quite acceptable signal-to-noise ratio.[7]

Peaks resulting from the formation of even-electron anions may be obtained if

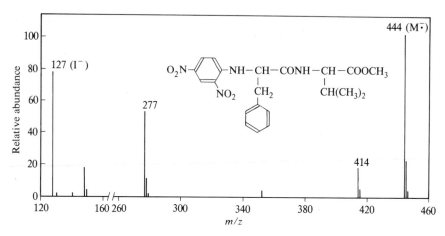

Figure 7-7

an appropriate concentration of methyl nitrite is introduced into the CI source (see Sec. 2-2D):

$$CH_3ONO + thermal\ "e" \rightarrow CH_3O^- + NO \qquad (7\text{-}4)$$

$$CH_3O^- + AH \rightarrow A^- + CH_3OH \qquad (7\text{-}5)$$

The anion A^- can only arise from the sample AH if CH_3O^- has been produced. This of course implies that the methyl nitrite has a comparable or

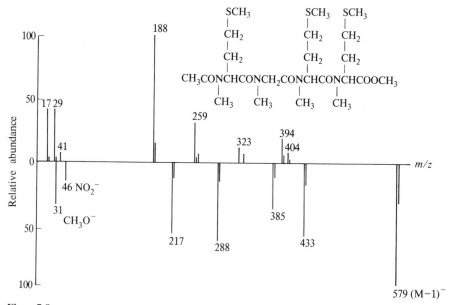

Figure 7-8

higher cross section for capture of thermal electrons than does the sample AH. Thus, this type of *Brönsted base spectrum* is available from samples which have not been modified to possess a high electron affinity. Since A^- is produced by true NICI involving the reaction of an anion and molecule [Eq. (7-5)] (and not an electron and a molecule), the sensitivities are less than those for high-pressure electron-capture spectra, and are comparable to PICI spectra.

By adding a small quantity of methyl nitrite to methane at 1 torr, it is possible to create a situation where both CH_3O^- and a population of excess thermal electrons exist in the CI source.[9] The simultaneous recording of positive- and negative-ion spectra by a quadrupole (Sec. 1-3B) can then be profitably employed. This technique is valuable when sample molecular weight and information from fragmentation are both required. As an example of this technique, the simultaneously recorded positive- and negative-ion mass spectra of the acylated permethylated peptide Met.Gly.Met.Met are reproduced in Fig. 7-8; the spectra were obtained with $CH_4 - CH_3ONO$ as the reagent gases.[9]

The PICI spectrum (upper trace) shows type **a** ions (m/z 188, 259, 404) and type **b** ions (m/z 323 and 394) which establish the sequence (Sec. 7-1). The NICI spectrum (lower trace) features a prominent $[M-H]^-$ ion, produced by reaction of the Brönsted base reactant ion, CH_3O^-, with the polypeptide sample. Electron capture produces a molecular anion radical (**7**, m/z 580), which may then dissociate into an alkyl radical and an anion (**8**) in which the negative charge is delocalized between the nitrogen and oxygen atoms of the amide linkage.

Ions of type **8** occur at m/z 217, 288, and 433 in the negative-ion trace of Fig. 7-8. They occur at m/z values 29 units (NCH_3) higher than the acyl ions of type **a** (Sec. 7-1) which are evident in the positive-ion trace.

In conclusion, it appears likely that applications of negative-ion mass spectrometry in high-pressure sources will grow, especially when sensitivity is of paramount importance and suitable electron-capture derivatives can be prepared conveniently.

7-3 FIELD IONIZATION (FI) AND FIELD DESORPTION (FD)

Two important features of FI and FD spectra are that (1) the molecular ions are produced with little excess energy compared with electron impact and (2) recorded fragment ions must be formed within ca. 10^{-11} s and therefore are usually of lower abundance than in EI spectra (Sec. 3-4).

If the absence of a molecular ion in an EI spectrum is suspected to be due to facile fragmentation of molecular ions which have been produced, then FI may be

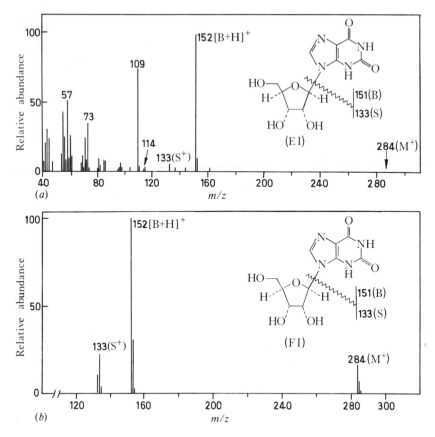

Figure 7-9

tried. However, in these circumstances, positive-ion chemical ionization would probably be the method of first choice. A comparison of this kind is made in Fig. 7-9, where the field ionization spectrum of the nucleoside xanthosine (b) is compared with the corresponding electron-impact spectrum (a).[10] Unmodified nucleosides do not yield abundant molecular ions in EI mass spectra, and the molecular ion of xanthosine is undetectable. The FI spectrum overcomes this problem: the molecular ion is readily detectable and the assignment of masses to base (B) and sugar (S) portions of the molecule is a simple matter. In fact, FI spectra of nucleosides are generally simple, and the molecular ion (or sometimes $[M+H]^+$), $[B+H]^+$, $[B+2H]^+$, and S^+ peaks are usually readily identifiable. Since most of the common base components have unique masses, FI spectroscopy offers a reliable means of identifying the base; but stereochemical features are inaccessible by this method. $[M+H]^+$ ions are common in FI mass spectra (see Fig. 7-9b) since ion-molecule reactions may occur in the dense layer close to the emitter surface where ionization takes place. However, these $[M+H]^+$ ions are easily distinguishable from M^+ ions if atomic compositions are determined; e.g., a

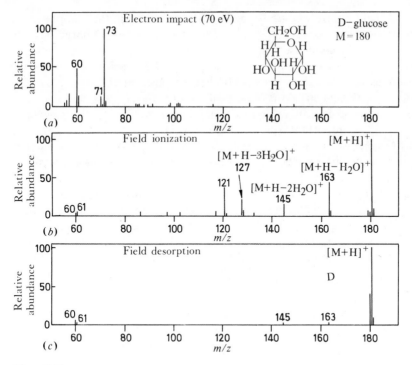

Figure 7-10

compound containing four nitrogen atoms must have an even-mass molecular weight.

Field desorption (FD) is a more important and more powerful technique since it can be applied to samples which are thermally involatile.[11] In part this is because electrical repulsion (rather than solely thermal energy) is used to destroy the interactions which hold the ionized sample molecule in its environment (Sec. 1-2D). The technique has advantages over FI (for unambiguous determination of molecular weight) even when sample volatility is significant. This is illustrated by the spectra of glucose which are reproduced in Fig. 7-10.

When loading an FD emitter (Secs. 1-2D and 2-3), the use of a microsyringe has the advantage that droplets of the sample solution can be placed accurately on the center of the wire. In favorable circumstances, as little as 10^{-11} g of sample can be detected by FDMS, although in general the technique is less sensitive than EI or CI. This is partly because in FI the ions leave the emitter over a wide angle compared with the very narrow angle of ions leaving the EI source. After a few hundred nanometers, the FI ions have acquired a kinetic energy of several kiloelectronvolts and divergent ion beams of this energy are difficult to focus into a narrow mass spectrometer slit.

In the case of a purely covalent sample, the FD spectrum is obtained after field ionization of the adsorbed sample at the tip of a microneedle of a conditioned wire. The small radius of curvature at such a tip allows the field strengths ($> 10^9 \, \text{V m}^{-1}$)

necessary for field ionization to be generated. However, if the sample is an electrolyte or zwitterion or contains salts or acids, ions of either charge sign are already present in the condensed phase. In such cases, the production of ions by field ionization of adsorbed sample is not necessary; the function of the external field is now to extract the existing positively charged ions from the surface layer. In these circumstances, field strengths of only 10^8 V m^{-1} are sufficient, and these can be generated at the surface of an *unconditioned* 10 μm wire, i.e., in the absence of needles.[12] This is important since untreated wire emitters can be used for obtaining FD spectra of nonelectrolytes if salts or acids are mixed with the sample. The sample may then be desorbed while solvating a proton or an alkali metal ion. In general, the addition of alkali metal salts (for example, NaI) to nonelectrolyte samples is preferred. Additionally, it is found that the admixture of substances of high viscosity in the liquid state (e.g., polyvinyl alcohol) improves the desorption behavior of many compounds. Thus, the FD mass spectrum of a mixture of the tetrasaccharide stachyose (**9**) (see also Fig. 7-3), polyvinyl alcohol, and NaI (the ratio of the weights is approximately $1:1:0.2$) shows m/z 689 $[M+Na]^+$ as the base peak. The spectrum was obtained with an untreated emitter, on to which the sample was loaded in methanol solution. The emitter was heated by passing a heating current of 25 mA through the wire.[12]

9

In FD work the distinction of possible M^+, $[M+H]^+$, and $[M+Na]^+$ ions is helped by doping the sample used for subsequent determinations with the corresponding lithium or potassium salts. The $[M+Na]^+$ species is often observed in spectra even when the sample has not been intentionally doped with a sodium salt; inadvertant contamination of samples with traces of Na$^+$ is common.

Since FD spectra contain so few abundant ions, mass marking of the spectrum is important. In this respect, a double-beam instrument is extremely useful. This instrument is essentially two mass spectrometers in one. Two ion beams are generated by adjacent sources, analyzed by common electric and magnetic sectors, and recorded by two adjacent electron multipliers. Since the output of the two mass spectra is on one ultraviolet chart, one spectrum can be used to mass mark the other. Thus, if one source is used to investigate an unknown when operating in the FD mode, the spectrum can be mass marked by using a reference compound in the other source—usually operating in the EI mode.

If a precise determination of ion mass is necessary in order to compute possible atomic compositions, then in favorable cases peak matching (Sec. 1-5) is possible. The reference peak may be provided by FI of a fluorocarbon compound introduced through a heated inlet system. Since FI of the reference compound is occurring at the same emitter as FD of the unknown, one problem is that the FI spectrum of the reference compound may be suppressed at the critical moment due to the spurt of ions being produced by the FD process. In some cases, this technique will fail either because the ion abundance due to the sample is too small or because the duration of the FD beam is too short. For these reasons, precise mass measurement of FD spectra may advantageously be carried out using a mass spectrometer which records ions in a focal plane (see Fig. 1-8); the ions are recorded with a photoplate, which is an integrating device.[13] Fluoroalkyl-phosphazenes have been recommended as high-molecular-weight reference compounds for FD mass spectrometry.[14]

Given that many compounds of great biological interest cannot be readily vaporized, even though their molecular weights lie in the range 300–1200 daltons, FD is an extremely important technique for the determination of molecular weights and molecular formulas of such compounds. Achievements with antibiotics[15] and other biologically important classes of compounds[16] have been reviewed.

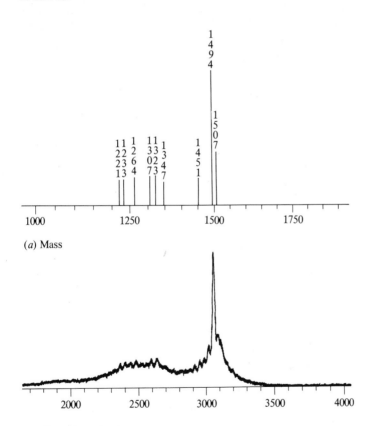

(a) Mass

(b) Channel number

Figure 7-11

7-4 CALIFORNIUM-252 PLASMA DESORPTION

This technique (Sec. 1-2E) must be regarded as the one to be employed when even FD fails to give molecular weight information. It has proved effective in handling highly polar molecules where other methods fail.[17] For example, it has been used to determine the molecular weights of antibiotics of the vancomycin group.[3,18,19] These compounds are based upon heptapeptide aglycones which bear between two and six additional sugar units; the molecular weights lie in the range 1200–2100 daltons.

The positive-ion spectrum of one member of the group, ristocetin A (**10**), gave a peak at m/z 2086 ± 5. After removal of five sugars by selective acid hydrolysis and diacetylation of the product, di-N-acetyl-ψ-aglycone (**11**) was obtained; O-methylation of this material with diazomethane gave **12**. The positive-ion spectra of **11** and **12** contained the most intense peaks at m/z 1410 ± 1 and 1494 ± 1, respectively; the latter data were obtained at a later date with better resolving power than that used for the study of **10**. Since the Na$^+$ ion abundance is relatively high in the spectra of these samples, the above peaks are likely to correspond to $[M+Na]^+$ species, suggesting molecular weights of 2063 ± 5, 1387 ± 1, and 1471 ± 1. These are the only direct determinations of molecular weight, and give confirmation to the structures **10**, **11**, and **12** (of molecular weights 2066, 1386, and 1470) which are based on chemical and NMR studies.

arabinosyl—mannosyl
|
10, R_1 = rhamnosyl—glucosyl—; $R_2 = R_3 = H$; R_4 = mannosyl
11, $R_1 = R_3 = R_4 = H$; $R_2 = CH_3CO$
12, $R_1 = R_3 = R_4 = CH_3$; $R_2 = CH_3CO$

The form of the original data for the spectrum of **12** is given in Fig. 7-11*b*, and the masses as determined and printed out by a computer program in Fig. 7-11*a*. Note that since most of the less-abundant peaks in the spectrum have not been interpreted, the spectra of a series of compounds (**10**, **11**, and **12**) are run before secure conclusions are derived.

Some molecules produce positive and negative molecular ions by ^{252}Cf PDMS, and others only positive molecular ions. In the case of **12**, the conclusion with regard to the molecular weight is reinforced by the presence of m/z 1470 ± 1 as the only abundant ion in the molecular ion region of the negative-ion spectrum.

REFERENCES

1. H. M. Fales, G. W. A. Milne, and R. S. Nicholson: *Anal. Chem.*, vol. 43, p. 1785, 1971.
2. R. C. Dougherty, J. D. Roberts, W. W. Binkley, O. S. Chizhov, V. I. Kadentsev, and A. A. Solv'yov: *J. Org. Chem.*, vol. 39, p. 451, 1974.
3. D. H. Williams, V. Rajanandra, and J. Kalman: *J. Chem. Soc. Perkin I*, p. 787, 1979.
4. H-W. Leung and A. G. Harrison: *Can. J. Chem.*, vol. 54, p. 3439, 1976.
5. F. Van Gaever, J. Monstry, and C. C. Van de Sande: *Org. Mass Spec.*, vol. 12, p. 200, 1977.
6. A. A. Kiryushkin, H. M. Fales, T. Axenrod, E. J. Gilbert, and G. W. A. Milne: *Org. Mass Spec.*, vol. 5, p. 19, 1971.
7. D. F. Hunt and F. W. Crow: *Anal. Chem.*, vol. 50, p. 1781, 1978.
8. R. C. Dougherty and C. R. Weisenberger: *J. Am. Chem. Soc.*, vol. 90, p. 6570, 1968; J. R. Chapman, D. Denne, and G. A. Errock: *Workshop on NICI*, National Institute of Environmental Health Sciences, Research Triangle Park, N.C., 1977.
9. D. F. Hunt, G. C. Stafford, Jr., F. W. Crow, and J. W. Russell: *Anal. Chem.*, vol. 48, p. 2098, 1976.
10. P. Brown, G. R. Pettit, and R. K. Robbins: *Org. Mass Spec.*, vol. 2, p. 521, 1969.
11. H. D. Beckey: *Int. J. Mass Spec. Ion Phys.*, vol. 2, p. 500, 1969; H. D. Beckey: "Principles of Field Ionisation and Field Desorption Mass Spectrometry," Pergamon Press, New York, 1978.
12. H. J. Heinen, U. Giessmann, and F. W. Röllgen: *Org. Mass Spec.*, vol. 12, p. 710, 1977.
13. H. R. Schulten and D. E. Games: *Biomed. Mass Spec.*, vol. 1, p. 120, 1974.
14. K. L. Olsen, K. L. Rinehart, Jr., and J. Carter Cook, Jr.: *Biomed. Mass Spec.*, vol. 4, p. 284, 1977.
15. K. L. Rinehart, Jr., J. Carter Cook, Jr., K. H. Maurer, and U. Rapp: *J. Antibiotics*, vol. 27, p. 1, 1974; K. L. Rinehart, Jr., J. C. Cook, Jr., H. Meng, K. L. Olsen, and R. C. Pandey: *Nature*, vol. 269, p. 832, 1977.
16. H. D. Beckey and H.-R. Schulten: *Angew. Chem. Int. Edn.*, vol. 14, p. 403, 1975.
17. R. D. MacFarlane and D. F. Torgerson: *Science*, vol. 191, p. 920, 1976.
18. W. J. McGahren, J. H. Martin, G. O. Morton, R. T. Hargreaves, R. A. Leese, F. M. Lovell, and G. A. Ellerstad: *J. Am. Chem. Soc.*, vol. 101, p. 2237, 1979.
19. Unpublished results from Professor R. D. MacFarlane's Laboratory.

EIGHT

ANALYSIS OF MIXTURES

8-1 INTRODUCTION

The commonest analytical problem to which mass spectrometry is currently applied is the identification of mixture components. This requirement for mixture analysis has increased enormously in the past few years and is largely a consequence of the proliferation of analytical problems in the environmental and health sciences. Improvements in instrumentation to meet this demand have meant that mixtures of widely differing complexity, concentration, and chemical nature can be successfully analyzed by an appropriate mass spectrometric technique.

One of the oldest and simplest methods used for identification of mixture components is fractional volatilization from the direct-insertion probe. Spectra of mixture components of different volatility can often be resolved, at least partially, by this method (see, for example, the derivatized peptide mixture discussed in Sec. 6-5D). This mode of separation is potentially applicable to all the ionization modes, and molecular ions of mixture components are often directly identified where field desorption and chemical ionization are employed.

Mixture analysis by mass spectrometry encompasses a variety of techniques; three of these are considered below. Two methods (GC/MS and LC/MS) involve chromatographic separation of the mixture components before ionization in the mass spectrometer. The main problem is the provision of a suitable interface between the chromatograph and the mass spectrometer. The third technique (collisional activation) involves the mass spectrometer alone: molecular ions of mixture components are selected individually in one sector of the instrument and caused to decompose prior to entering another sector for analysis. The chapter is concluded by a discussion of some of the procedures required to obtain quantitative results from mass spectral measurements (particularly from GC/MS).

8-2 GAS CHROMATOGRAPHY—MASS SPECTROMETRY (GC/MS)

A. The Hardware

The separation and detection of components from a mixture of organic compounds is readily achievable by gas chromatography. Furthermore, limited characterization of unknown components is often possible from retention times appropriate to the particular column used. Mass spectrometry, because of its high sensitivity and fast scan speeds, is the technique most suited to provide definite structural information from the small quantities of material eluted from a gas chromatograph. The association of the two techniques has therefore provided a powerful means of structure identification for the components of natural and synthetic organic mixtures. Mass spectra of acceptable quality are potentially obtainable for every component that may be separated by the gas chromatograph, even though the components may be present in nanogram quantities and eluted

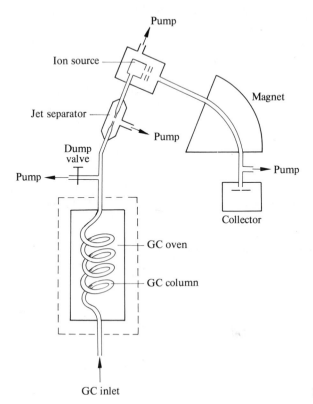

Figure 8-1

over periods of only a few seconds. Consequently, GC/MS is a widely used analytical tool whose fields of application are constantly expanding.[1,2]

A schematic diagram of a GC/MS system is shown in Fig. 8-1. Typical GC/MS outputs of total ion current versus time (representing the quantities of individual components eluted from the GC column) can be seen in Figs. 8-3 and 9-4. The particular system illustrated in Fig. 8-1 incorporates a packed column, jet separator, and magnetic mass spectrometer. However, different types of column, interface, and spectrometer are in common use for GC/MS analysis, and the basic requirements and functions of these various units are considered below.

The *gas chromatograph* must be capable of efficient separation of mixture components, and columns should be easily interchangeable. The choice of column phase is governed not only by its efficiency in separating the types of compound injected but also by the extent of column "bleed". Excessive elution of column material, which occurs particularly at high temperatures, tends to block the separator and produces a high background spectrum. Under such circumstances, it is difficult to identify the spectra of minor components above background although computer routines may alleviate this problem (see Sec. 9-3). Several parameters should be controlled in order to minimize column bleed. Columns should (1) be made from low bleed phases, (2) have as low a loading of stationary phase as possible, (3) be well conditioned, and (4) be used at the lowest possible temperature. Stationary phases will bleed excessively if they are improperly coated on the support. Among the stationary phases used for GC/MS analysis, silicone phases are particularly recommended (for example, OV17, OV101, SP-2100, SP-2250, SP-2300).

Another critical parameter in GC/MS analysis is the gas flow rate through the GC column. Although helium flow rates of up to $60 \, \text{ml min}^{-1}$ are achievable with packed columns in GC/MS, the transfer efficiency of compounds across the GC/MS interface is reduced at these high rates. If glass capillary or SCOT columns are used (usually without a separator), carrier-gas flow rates into the ion source may be up to $10 \, \text{ml min}^{-1}$. The consequence of these limitations on flow rates is that occasionally the optimum rate for GC analysis alone must be reduced for GC/MS analysis. Incidentally, the ion current due to He^+ produced in the ion source may be eliminated by employing an electron energy of 20 eV (below the ionization energy of He). However, this refinement is not usually necessary.

Although glass capillary columns can achieve highly efficient separation, their use in GC/MS analysis has until recently been limited. Reasons for not previously employing capillary columns have included (1) the problems of interfacing with the mass spectrometer, (2) the lack of commercial availability, (3) the fact that mass spectrometer scan speeds were too slow to accommodate the sharp peaks eluted from capillary columns (often less than 3 s wide), and (4) the low sensitivity on some instruments. However, most of these reasons no longer apply and capillary-column GC/MS is used widely in analytical chemistry and biochemistry.

In order to minimize source contamination it is strongly advisable to establish the correct GC conditions using an appropriate GC detector (for example, a flame ionization detector, FID) *before* coupling the column (packed or capillary) to the mass spectrometer. In order to achieve this aim more efficiently, some GC/MS systems incorporate adjacent ports for coupling to the GC column outlet, one leading to an FID and the other to the mass spectrometer interface. It is then a simple matter to secure the column to the mass spectrometer interface after GC conditions have been established.

Another important requirement in GC/MS is effective temperature programming of the GC column. In many modern GC/MS systems this parameter is under microprocessor control, along with gas flow rates, injection and separator temperatures, as well as the mass spectrometer scan parameters.

The *interface* between the GC and MS components should be capable of sustaining a large pressure drop from about 1 atm in the gas chromatograph to below 10^{-6} torr in the ion source (except where chemical ionization conditions are employed). It is therefore essential to remove most of the carrier gas from the sample effluent. The transfer of sample across the interface should be rapid and efficient. Two types of interface have found widespread use between glass capillary columns and the mass spectrometer.[3] One technique involves a direct connection by means of a platinum capillary. The other technique is the open-split connection in which the inlet line to the mass spectrometer is a flow resistance, from which excess gas escapes to the atmosphere.

Several types of interface, or *separator*, have been developed commercially for use with packed columns. The membrane molecular separator has been used with some success and depends upon the preferential diffusion of organic molecules through a methyl silicone membrane and rejection of most of the carrier gas. However, its upper temperature limit ($\sim 220°C$) restricts its usefulness.

The commonest GC/MS interface is the all-glass jet *separator* (Fig. 8-2). Most of the carrier gas is eliminated by utilizing the relatively fast diffusion of the low-molecular-weight carrier gas into a pumped interspace between two aligned orifices (typically of diameter 50–100 μm). This separator meets the important

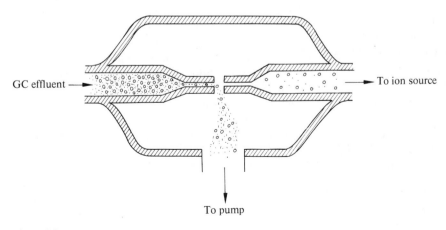

GC effluent → ... → To ion source

To pump

Figure 8-2

requirements of having no dead volume and a very small surface for transference of the sample. Its efficiency is high and up to 50 percent of the sample is transferred across the jet when high-speed pumping and the narrowest practicable orifice are employed. Analytically, the jet separator is particularly advantageous in the handling of high-molecular-weight samples since it can operate at temperatures up to 300°C. One drawback is its inflexibility for widely differing flow rates of carrier gas.

The narrow constriction in the jet separator renders it susceptible to obstruction by column effluent. Provision of a sinter between the column and separator can partly alleviate this problem. It is good practice to inspect the jet regularly, preferably under magnification, for possible contamination. To facilitate cleaning it is advantageous to have a detachable jet, secured at ground glass joints. Several procedures may be adopted to clean a jet, including (1) use of a fine wire to dislodge particulate matter, (2) immersion in organic and/or caustic solvents, and (3) heating in a glass-blower's oven just below the melting point of the glass. Transfer efficiency and separator lifetime are usually improved by treatment with an appropriate silanizing reagent.

The need to avoid contamination of the separator and mass spectrometer should be an everpresent consideration during GC/MS operation. With this purpose in mind a *dump-valve* and/or a *splitter* may be incorporated into the line between the GC and separator. The dump-valve is operated when required during the GC/MS run and its function is to vent unwanted GC effluent (for example, major components of known structure) to the atmosphere. A splitter diverts part ($\sim$ 5 percent) of the column effluent to a detector (usually an FID). This device is useful in monitoring the GC trace when utilizing the dump-valve. However, its presence is not essential in this respect and the difficulties encountered in operating a splitter may outweigh its advantages.

A major requirement for a *mass spectrometer* used on-line with a gas chromatograph is a capability for fast scanning. The time of elution of a gas chromatographic peak normally varies from about 1 s to about 1 min. Hence, to avoid distorted mass spectra, minimum total scan periods of less than a second are required. Appropriate rates are obtainable both with quadrupole and magnetic-sector instruments.

The dynamic range of the mass spectrometer for peak recognition should be as high as possible, especially where minor components of mixtures have to be analyzed. The mass spectrometer should also carry a total ionization monitor which acts as a detector for the gas chromatograph.

There are several modes of operation of the mass spectrometer to be considered when setting up a GC/MS system. If sensitive detection work is to be carried out, a unit for multiple-ion monitoring is essential (see Sec. 8-2C). If high-resolution work is to be performed in conjunction with a computer, a double-focusing magnetic spectrometer must be chosen. However, exact mass measurements may be obtained from quadrupole spectrometers under microprocessor control (see Sec. 9-2F). It is also an advantage to possess a facility for different modes of ionization. Both positive- and negative-ion chemical ionization have been highly successful in GC/MS work and ion sources are available which can switch from CI to EI operation during the elution of one GC peak. Field ionization (FI) GC/MS has also been used, but with limited success. Molecular ion peaks of reasonable intensity are usually attainable, but efficient passage of the GC components across the FI wire is a design feature that is not always achieved.

The discussion above has enumerated the requirements for a successful GC/ MS system. However, in order to digest, manipulate, and interpret the vast amounts of data obtained, a computerized data system is an essential addition (see Sec. 9-3).

B. Applications

The applications of GC/MS are numerous and frequent reference will be made to this powerful combination later in the book. Below, in Sec. 8-2C, selected ion monitoring (SIM) will be considered as an important GC/MS technique, and the application of SIM to quantitative measurements is covered in Sec. 8-5. In Chap. 10, various analytical applications of GC/MS in biomedical, environmental, and industrial research will be found, and the use of computerized data systems for GC/MS analysis is considered separately in Sec. 9-3.

In this section, therefore, the discussion of GC/MS applications is confined merely to an outline of some derivatization procedures employed for preparation of samples, followed by one illustrative example of an analysis.

Many classes of compounds need to be derivatized before gas chromatographic separation, either because of involatility or thermal instability. For mass spectrometric identification it is also desirable that the spectra of these derivatives should include molecular ions and/or structurally significant fragment ions. Where low levels of compounds are being investigated it is preferable that the derivatization reaction should proceed with a high yield.

A useful derivative for alcohols is the TMS ether (see Fig. 8-3, Table 10-1, and the accompanying discussions). These derivatives are volatile, have low polarity, and give rise to characteristic fragmentations, usually via α cleavage. However, TMS ethers, or even the t-butyldimethylsilyl (TBDMS) ethers preferred by some workers, are not always the best choice from a mass spectrometric viewpoint. For example, free steroids yield spectra that are more structure-specific than their corresponding TMS derivatives. If low volatility necessitates derivatization, in most cases the preparation of keto compounds from hydroxy steroids is preferable to TMS ether formation (see Table 10-1). Carboxylic acids (for example, **1**) can also yield TMS derivatives, although it is usual to employ the methyl ester (see Fig. 9-4).

Gallic acid, **1**

Alternative derivatives for alcohols are acetyl esters and methyl ethers. The overriding factor in the choice of derivative is normally a chromatographic one, so that efficient separation can be achieved before mass spectrometric identification. However, the derivative should not undergo facile fragmentations in the mass spectrometer with charge retention on the derivative group to such an extent that the spectrum is dominated by structurally noninformative ions. For example, in some acetates and TMS ethers, m/z 43 (CH_3CO^+) and m/z 73 (Me_3Si^+), respectively, may carry a high percentage of the ion current. Permethylation is a particularly favorable alternative method in this respect (see Sec. 6-5D).

O-Methyloximes have proved to be suitable derivatives for aldehydes and ketones despite the complication of *syn-* and *anti-*isomers. These derivatives yield stable molecular ions and structurally informative fragmentation patterns.

Negative-ion GC/MS sometimes requires special derivatization to increase the electron affinity and therefore facilitate ionization by electron capture. In particular, CF_3CO- and C_2F_5CO- derivatives are commonly used for alcohols. Derivatives which incorporate the *p*-nitrophenyl function are excellent for electron capture, but may render the compound too involatile for gas chromatography.

An example of the identification of mixture components by GC/MS is illustrated in Fig. 8-3. A procedure was devised for determining the constituents of ink from ancient handwritten parchments[4] in which (1) 1 mg of ink was detached by a scalpel from various letters on different pages, (2) the ink was hydrolyzed by hydrochloric acid to yield a mixture of monosaccharides and tannic acids, (3) after evaporation to dryness, TMS derivatives were formed by reaction with a hexamethyldisilazane–TMS chloride–pyridine mixture (2:1:10) which is effective for both sugars and tannic acids, and (4) 1 μl of the mixture was injected into a packed SE-30 column.

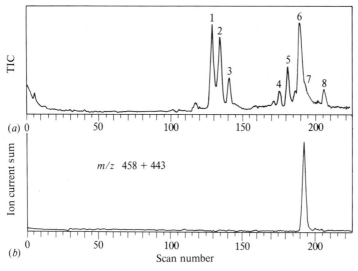

Figure 8-3

Table 8-1 Compounds identified from an ancient ink hydrolyzate (see Fig. 8-3)

Peak no.	Compound	Peak no.	Compound
1	β-arabinopyranose	5	α-glucopyranose
2	α-arabinopyranose	6	β-galactopyranose
3	β-arabinofuranose	7	gallic acid
4	α-galactopyranose	8	β-glucopyranose

The total ion current (TIC) trace obtained is depicted as a function of time in Fig. 8-3a. By comparison with the mass spectra and GC retention times of standard components, compounds 1 to 6 and 8 were identified as TMS derivatives of monosaccharides (see Table 8-1). Acid hydrolysis and TMS-ether formation cause mutarotation of sugars so that both furanoside and pyranoside forms of α and β anomers were identified. Peak 7, which appears as a shoulder on peak 6, was suspected to be the TMS derivative of gallic acid (**1**) and a computer-reconstructed trace (Fig. 8-3b) of the ion-abundance sum of M^+ and $M^+ - CH_3$ from this compound (at m/z 458 and 443, respectively) strikingly supports this structure. Such ion-abundance traces will be considered in Sec. 9-3.

The above example illustrates how GC/MS analysis can identify mixture components at low levels (the nanogram level for individual components in this example). Structural identification of ink components from various manuscripts revealed differences from various geographical locations and suggested methods for preservation and restoration of ancient manuscripts.

C. Multiple-Ion Detection

An important analytical application of the GC/MS combination is the identification and/or quantitative assay of compounds eluting from a GC column by single- or multiple-ion monitoring (SIM and MIM, respectively). These techniques (employed with both magnetic or quadrupole instruments) use the mass spectrometer as a sensitive but selective GC detector.[5] If an ion of given m/z value is passed continuously through the collector slit or if electronic circuitry (often under microprocessor control) is employed to switch rapidly between a series of prominent ions, little of the total ion current is wasted. This contrasts with the scanning of a full mass spectrum where on average a high percentage of the total ions are colliding with the walls, slits, etc. Consequently, greater sensitivity is achieved with SIM and MIM. For example, an instrument which is capable of giving a full mass spectrum from 1 ng of sample might detect 1 pg (or lower) from the same compound by SIM. At such low levels SIM and MIM are usually employed to detect or quantitate expected compounds rather than identify unknown compounds. Nevertheless, a meaningful partial mass spectrum may be extracted from a relatively high background (due to column bleed, etc.) if a sufficient number of ions are monitored in MIM. Any number from 1 to 20 ions

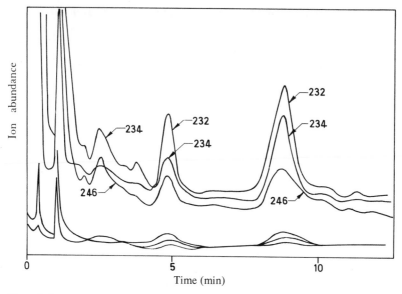

Figure 8-4

may be appropriate to a specific problem. The higher the number the less sensitive and more selective the detection method becomes, and the user should balance these two factors according to the problem encountered.

An application of MIM to the identification in human blood of metabolites of the drug chlorpromazine (**2**) is illustrated in Fig. 8-4.[6] The mass spectra of chlorpromazine and its side-chain derivatives yield abundant ions at m/z values of 246 (**3**), 232 (**4**), and 234 (a ^{37}Cl and, to a lesser extent, a ^{34}S isotope peak of m/z 232). These three ions and others were monitored continuously from the GC effluent of the derivatized blood extract (see Fig. 8-4). The abundance of the three ions rose and fell simultaneously for two of the eluted compounds. These elutants are therefore strongly suggested to be chlorpromazine side-chain derivatives, and the structures were confirmed by monitoring suspected molecular ions and utilizing authentic GC retention times. The two derivatives were in fact the trifluoroacetates of des- and didesmethylchlorpromazine.

For quadrupole mass filters, multiple-ion monitoring is a relatively simple and commonplace procedure. The voltages on the rods are low and can be switched rapidly between those necessary to focus chosen ions. Ions can be monitored over a wide range and computer-controlled MIM is now widely applied to quadrupole GC/MS systems. In a magnetic-sector instrument the ion intensity of two or more mass numbers can be recorded within a short time interval by use of an accelerating voltage alternator. In a single-focusing instrument, ions of different mass numbers can be alternately brought into focus at the collector by changing the accelerating voltage in the ratio of their masses. In a double-focusing instrument the same function is performed while maintaining a constant ratio between the accelerating voltage and the electrostatic analyzer voltage. If only two masses are monitored the above operation can be performed on the standard peak matching unit of a double-focusing instrument. When more than two ions are to be monitored, appropriate modules (often under microprocessor control) are available to monitor as many as 20 ions in subsecond total scan times.

The application of double-focusing magnetic instruments to multiple-ion detection is also important in that it enables high mass resolution to be utilized. When ions of the same nominal m/z value are monitored, it is evident that results could be compromised where more than one elemental composition (and therefore exact masses) is possible. High-resolution SIM might alleviate this problem if ion intensities are monitored within narrow mass ranges (e.g., 0.005 mass units or less). Since focusing slits are partially closed to achieve high mass resolution, the gain in specificity is accompanied by a loss in sensitivity. A typical improvement in specificity upon increasing the resolving power (RP) is shown in Fig. 8-5.[7] The m/z value 174.1 was monitored in a urine extract; the two resolving powers illustrated are 1000 and 10,000.

It is evident from the above example that SIM and MIM are sometimes nonspecific. Various alternatives are available to increase specificity, including:

1. High resolution (as discussed above).
2. The selection of a greater number of ions to be monitored.

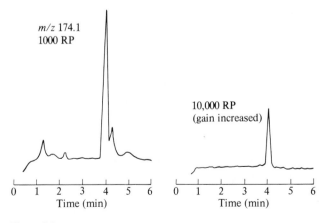

Figure 8-5

3. The use of higher mass peaks by appropriate derivatization: peaks at lower masses are potentially produced from fragmentation of more than one molecular ion.
4. Chemical ionization. Less fragmentation occurs.
5. Negative ionization. This technique is highly sensitive for appropriate compounds and the extent of fragmentation is usually low. However, the range of compounds is limited.

The MIM technique may, of course, be applied directly to involatile probe samples, avoiding the use of gas or liquid chromatography. The danger of nonselectivity is high, but reliable results can be achieved by careful probe-temperature control.

It is pertinent at this stage to draw a distinction between chromatograms obtained by SIM or MIM (for example, Figs. 8-4 and 8-5) and those obtained by computer reconstruction of ion intensities (for example, Fig. 8-3). The former traces are highly sensitive and are a result of rapid switching between selected masses. The latter traces are usually obtained from retrospective selection of individual ion intensities from whole spectra and are therefore less sensitive.

8-3 LIQUID CHROMATOGRAPHY–MASS SPECTROMETRY (LC/MS)

High-performance liquid chromatography (HPLC) has been developed as a separation technique such that its separation capability and analysis time are comparable to those attainable via gas chromatography (GC). Furthermore, HPLC can handle the many polar and thermally labile compounds not amenable to GC. It is this versatility that has led to the exploration of on-line coupling of the HPLC to the mass spectrometer (LC/MS). As with GC/MS in its infancy, several interfaces for LC/MS have been investigated. The next few years will reveal which of these interfaces will become the most favored (compare the jet separator in GC/MS).

This section covers the current status of LC/MS as a viable system for detection or structural analysis of the components of organic mixtures. The potentials and drawbacks of different types of interface are briefly discussed, together with an example of an application of the technique.

A. The Hardware

Some of the requirements for a successful LC/MS interface are as follows: (1) the solute should be transferred efficiently across the interface (for example, > 30 per cent), (2) the solvent must be removed efficiently by the interface (unless the solvent is acting as the ionizing reagent for chemical ionization), and (3) chromatographic separation must not be impaired by the interface. The types of interface employed include the following: (1) a moving wire, (2) a moving belt, (3)

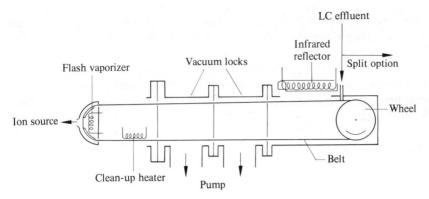

Figure 8-6

a fine-glass capillary (with a chemical ionization or atmospheric-pressure ionization source), (4) a jet separator (with a micro-LC and a chemical ionization source), and (5) a silicone membrane separator. In this section, two interfaces will be considered, one applicable to electron impact and the other to chemical ionization mass spectrometry.

The *moving-belt* LC/MS interface[8] is illustrated in Fig. 8-6. The effluent (liquid solvent plus solute) from the LC is deposited on the belt which moves through two vacuum locks where the solvent evaporates. The remaining solute passes alongside a flash vaporizer and is volatilized into the ion source. Additional features shown in Fig. 8-6 include an infrared reflector preceding the vacuum locks and a "clean-up" heater following the vaporizer. The infrared reflector enhances the evaporation of polar solvents (for example, methanol, acetonitrile, water) and thus improves the solvent capacity. The "clean-up" heater helps to remove residual sample which would cause a "recycle" or "ghost" peak on the next pass of the belt. However, source contamination is a problem which is accentuated in all types of LC/MS separators by the use of inorganic buffers.

Volatile solvents such as hexane can be transmitted by the interface at rates greater than $1\,\text{ml}\,\text{min}^{-1}$, but for polar solvents the maximum rate is somewhat lower. Better than 30 percent transfer efficiency can be achieved and samples can be handled at the nanogram level. The main disadvantages of the interface is peak recycling with accompanying source contamination. Despite these problems the moving belt has been used not only as an LC/MS interface but also in place of a direct insertion probe for obtaining mass spectra from a sample dissolved in a solvent.

Various LC/MS *capillary* interfaces have been designed for use with a chemical ionization (CI) source, in which the LC solvent becomes the CI reaction medium. A major disadvantage of this type of interface lies in finding an appropriate LC solvent which will also act as a CI reagent gas for the classes of compounds to be analyzed.

An LC/MS capillary interface has been described[9] which inserts directly into a CI source of an unmodified quadrupole mass spectrometer. The problems associated with magnetic-sector instruments (for example, high-voltage sparking and electrical conductance through the capillary) are therefore avoided. The LC column eluant is split so that a small portion is passed through the capillary (in this case 25 cm long and 75 μm internal diameter, narrowing to 5 μm at the tip) directly into the source. Among solvents utilized in this system are acetonitrile/water (60/40) and methanol/pentane (25/75). Efficient volatilization of the solvent is achieved by maintaining the ion source temperature at ~ 250°C, and full mass spectra are obtainable from samples in the nanogram range. LC flow rates greater than 2 ml min⁻¹ are avoided.

B. Applications

The LC/MS combination is potentially applicable to a wide range of underivatized mixtures. The principle difficulty lies in the identification of an appropriate solvent–interface–ionization-method combination for a specific analytical problem.

Figure 8-7 illustrates the application of LC/MS to a pesticide mixture analysis using the capillary interface described above.[9] The trace of TIC versus time for a mixture of the pesticides difonate (**5**), parathion (**6**), and methyl parathion (**7**) is shown in Fig. 8-7a. A 25/75 methanol/pentane mixture was used as the LC eluant and CI reagent gas. Abundant protonated molecular ions were generated under these conditions, and the selectivity and sensitivity of the method are illustrated in Fig. 8-7b. Dual ion traces of the m/z values for the MH⁺ ions from **5** and **7** were obtained as shown from an injection containing 500 pg of each component.

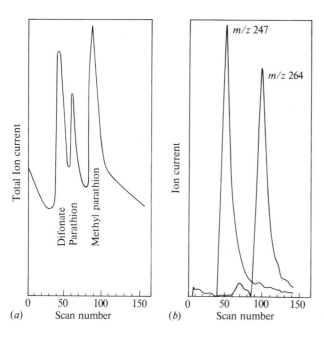

Figure 8-7

Acceptable signal-to-noise ratios would clearly be obtainable for quantities lower than this.

Difonate, **5**

Parathion, **6**

Methyl parathion, **7**

8-4 COLLISIONAL ACTIVATION

The principle of mixture analysis by collisional activation in a reversed geometry instrument is simple. Individual molecular ions of a given mass are selected by the magnetic analyzer (MA) and are then passed through a "collision cell" containing an inert gas at a pressure of 10^{-5} to 10^{-4} torr. Excitation of the molecular ions is achieved by conversion of translational (kinetic) energy into internal energy. The resultant fragment ions are characteristic of the molecular ion structure. These ions, which possess kinetic energy in proportion to their masses (see Sec. 3-2B), are analyzed by scanning the electrostatic analyzer (ESA) voltage. Each molecular ion can be selected in turn by the MA and its collisional activation (CA) spectrum, obtained as indicated above, is used to characterize its structure. The technique is outlined in Fig. 8-8. A hypothetical mixture of molecules M_1, M_2, and M_3 is ionized after evaporation from the direct-insertion probe. In this example the

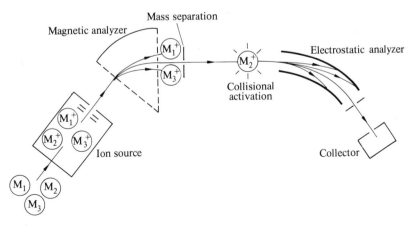

Figure 8-8

molecular ion M_2^+ is selected by the MA and its collisional activation spectrum is determined by scanning the ESA voltage and sweeping successive fragment ions through the collector slit.

The obvious advantage of collisional activation for mixture analysis is that no chromatographic separation of individual components is required prior to mass spectrometric analysis. However, there are several disadvantages of the technique.

The first drawback is the possible lack of selectivity for molecular ions passing through the magnet (particularly following electron-impact ionization). In other words, a molecular ion at a given m/z value may have the same nominal mass as one formed by degradation of one or more different molecular ions. This problem of lack of selectivity, which is particularly prevalent at low m/z values, is discussed in Sec. 8-2C. However, it should be emphasized that isomeric ions are, of course, inseparable by the magnetic analyzer and none of the techniques discussed in Sec. 8-2C to improve selectivity are applicable to such ions. The second disadvantage is one of sensitivity. Only a fraction of the total ion current passes through the exit slit from the magnetic analyzer when an individual ion is being selected from a complex mixture. Furthermore, the total ion current is somewhat diminished by passage through the collision cell.

A third disadvantage associated with this method of mixture analysis is that some knowledge about the expected components is required prior to analysis.

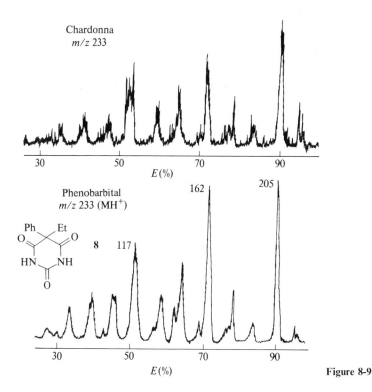

Figure 8-9

Otherwise the user of the technique must estimate which ions to select for collisional activation.

Despite the problems described above, some impressive results have been obtained for mixture analysis by collisional activation.[10] Crude mixtures have been analyzed directly and Fig. 8-9 illustrates the identification of phenobarbital (**8**) in a pharmaceutical preparation (Chardonna). The preparation was placed directly on the probe without any pretreatment. The collisional activation spectrum of m/z 233 (the protonated molecular ion formed by methane chemical ionization) is very similar to that formed from authentic phenobarbital. Collisional activation has also been used to analyze components of peptide mixtures ionized by OH$^-$ [11]

It can be seen from the collisional activation spectra shown in Fig. 8-9 that the fragmentation peaks are broad, due to translational energy release during the fragmentation processes (see Sec. 3-2D). In fact the effective mass resolution is frequently ca. 100 in such spectra; this resolution is insufficient for adequate separation of adjacent peaks. However, if collisional activation spectra are obtained via appropriate linked scans (see Sec. 3-2B), this problem can be alleviated. For example, the spectra acquired from B/E linked scans on either standard or reversed-geometry instruments exhibit peaks with effective mass resolution usually between 200 and 1000.

8-5 QUANTITATIVE MASS SPECTROMETRY

Throughout the 1960s organic mass spectrometry underwent rapid expansion and became recognized as a sensitive technique for structural elucidation, providing information complementary to that obtained by other spectroscopic methods. Since this period of growth, many users of mass spectrometry, particularly in the fields of biochemistry, pharmacology, clinical chemistry, and environmental studies have developed the quantitative aspects of the technique. Problems from these various disciplines can be approached by broadly similar mass spectrometric procedures and accurate quantitative measurements (better than 10 percent precision) are achievable at the picogram level.

Since quantitative measurements employing mass spectrometry are invariably performed on mixtures, this topic is conveniently covered here. More detailed accounts of this rapidly expanding field may be found elsewhere.[7]

Reliable quantitative measurements in mass spectrometry require a suitable internal standard. Quantitative analysis is usually performed via GC/MS; consequently, derivatization of the sample and standard is frequently required. One (or several) characteristic ions in the sample and standard are selected and continuously monitored (see Sec. 8-2C above). The signal intensity from the sample is related to that from a known amount of standard, preferably after construction of a. calibration curve over the approximate quantity range of the unknown compound to be measured. In order to obtain reliable results it is necessary to (1) perform repeat experiments and (2) run "blanks" to determine background signals at the masses being monitored.

Table 8-2 Examples of standards for quantitative mass spectrometry

Compound name	Structural formula	Standard	Type of standard
Methyl parathion	O_2N—⟨benzene⟩—$OP(OMe)_2$ (S double bond)	O_2N—⟨benzene⟩—$OP(OCD_3)_2$ (S double bond)	Isotopically labeled analog
Diphenyl hydantoin	Ph Ph hydantoin ring, O, O, HN, NMe, C=O	Ph Ph hydantoin ring with * (¹³C enriched)	As above
Indole 3-acetic acid	indole—CH_2CO_2H, N–H	methyl-indole—$CH_2\ CO_2H$, N–H, H_3C, CH_3	Homolog
Phenyl pyruvic acid	⟨phenyl⟩—CH_2COCO_2H	⟨tolyl, CH_3⟩—$COCO_2H$	Isomer
Imipramine	dibenzazepine, N–$(CH_2)_3NMe_2$	phenothiazine (S), N–$(CH_2)_3NMe_2$ (promazine)	Similar class

* Carbon atom enriched with ^{13}C.

Four types of standard compound are usually considered for quantitative measurements, namely: (1) an isotopically labeled analog, (2) a close homolog, (3) an isomer, and (4) a compound of the same (or similar) chemical class. Examples of each of these are shown in Table 8-2. These examples are merely illustrative and the choice of a particular standard does not imply that it is uniquely suitable.

There are several advantages associated with the use of isotopically labeled standards, namely: (1) the sample and standard have almost the same GC retention times and therefore errors due to variation of GC/MS parameters with time are minimized, (2) an excess of the standard may be used to transmit very small amounts of the sample through separation and derivatization steps, and (3) the sample and standard have very similar physical properties in terms of derivatization rates and partition coefficients. This means that extraction and derivatization conditions need not be so stringently controlled. Although *stable*

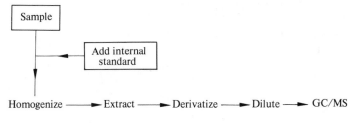

Figure 8-10

isotopes are normally employed it is sometimes preferable to add a small quantity of radioactive tracer to facilitate extraction.

The advantage of the other standards shown in Table 8-2 is that normally no special syntheses are required. Where an isomeric ion from a suitably fragmenting homolog is employed only one ion need be monitored, but it is of course essential that the GC column separates the standard and sample. As a general rule it is often recommended that a close homolog be considered first, as this is usually the cheapest route. Economic factors and the accuracy of the assay required are usually the most important considerations.

When performing quantitative measurements on crude extracts using GC/MS techniques it is usually advantageous to add the internal standard at an early stage of the extraction/derivatization procedure. In most cases the standard is added before separation and derivatization (for example, to a urine or plasma sample) so that both the sample and standard undergo identical processes before GC/MS analysis. A typical assay procedure is shown in Fig. 8-10.

One common difficulty in quantitative mass spectrometry is the lack of specificity due to the possible multiplicity of peaks at the nominal mass(es) being monitored. This problem may be overcome by employing high-mass derivatives (for example, TMS or fluorinated derivatives) where the probability of the presence of more than one peak is reduced. Alternatively, high resolution may be used (for example, 5000–10,000 resolving power) to separate overlapping peaks. A loss of sensitivity occurs at higher resolution and a balance must be struck between a gain in specificity and a loss of sensitivity (see also Sec. 8-2C).

Although quantitative measurements are performed in most cases using electron-impact (EI) ionization, an increasing number of assays employ chemical ionization (CI), sometimes negative ionization (NI), and even field desorption (FD). These techniques are more selective since they usually afford simpler spectra containing characteristic ions in the molecular ion region. For example, C_2F_5CO-derivatives are employed to assay biogenic amine metabolites by negative CI. FD has limitations as a quantitative method because of its lack of reproducibility, but it has been used to assay polar compounds which are not amenable to EI or CI. Chromatography is not always necessary for assay by FD and a controlled increase of emitter current can effect partial separation of mixture components. Even tissue homogenates have been assayed by such techniques.

The majority of quantitative measurements by mass spectrometry are carried

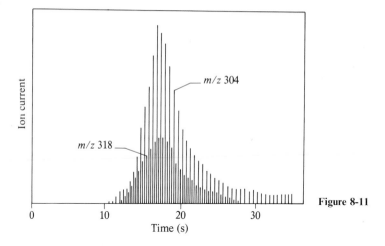

Figure 8-11

out using a GC inlet system as indicated above. However, reliable quantitation can be achieved using the direct-insertion probe via the integrated ion current (IIC) technique. Most of the applications to date have been directed to components of complex biological mixtures. One major ion is chosen for the compound to be assayed together with one from a standard. They are both monitored as the probe temperature is raised under carefully controlled conditions. The usual precautions should be adopted to maintain selectivity (see Sec. 8-2C).

The output obtained from the IIC technique is a series of peak profiles, each corresponding to one particular ion. Quantitation is achieved by measuring areas under each peak in comparison to the standard. The type of curve obtained is illustrated in Fig. 8-11 for the dansyl derivatives of pyrrolidine (**9**) and piperidine (**10**).[12] The peak profiles shown are relatively simple but where complex biological mixtures are employed their shapes are often complicated, possessing several maxima.

9, M = 304   **10**, M = 318   Dansyl

REFERENCES

1. C. J. W. Brooks and B. S. Middleditch: In R. A. W. Johnstone (ed.), "Specialist Periodical Reports," vol. 5, The Chemical Society, London, 1979.

2. A. L. Burlingame, C. H. L. Shackleton, I. Howe, and C. S. Chizov: *Anal. Chem.*, vol. 50, p. 346R, 1978.
3. H.-J. Stan and B. Abraham: *Anal. Chem.*, vol. 50, p. 2161, 1978.
4. P. Arpino, J.-P. Moreau, C. Oruezabal, and F. Flieder: *J. Chromatog.*, vol. 134, p. 433, 1977.
5. C. Fenselau: *Anal. Chem.*, vol. 49, p. 563A, 1977.
6. C.-G. Hammer, B. Holmstedt, and R. Ryhage: *Anal. Biochem.*, vol. 25, p. 532, 1968.
7. B. J. Millard: "Quantitative Mass Spectrometry," Heyden and Son, 1978.
8. W. H. McFadden, H. L. Schwartz, and S. J. Evans: *J. Chromatog.*, vol. 122, p. 389, 1976.
9. J. D. Henion: *Anal. Chem.*, vol. 50, p. 1687, 1978.
10. R. W. Kondrat and R. G. Cooks: *Anal. Chem.*, vol. 50, p. 81A, 1978; F. W. McLafferty and F. M. Bockhoff, *Anal. Chem.*, vol. 50, p. 69, 1978.
11. C. V. Bradley, I. Howe, and J. H. Beynon: *J.C.S. Chem. Comm.*, p. 562, 1980.
12. N. Seiler and B. Knödgen: *Org. Mass Spec.*, vol. 7, p. 97, 1973.

NINE

COMPUTER TECHNIQUES

9-1 INTRODUCTION

It is evident from the foregoing chapters that mass spectrometry has become a powerful tool for assisting the structure elucidation of organic molecules. The products of organic ionic reactions, occurring in a few microseconds or less, are recorded in periods varying from less than a second to several minutes. The resulting mass spectrum is then ready for analysis by a human interpreter and this process may take a considerable time, depending on the skill and experience of the interpreter. If the compound is likely to have a known mass spectrum, reference can be made to compilations of spectra. Further details such as the exact mass and elemental composition of the molecular ion (or other ions) require high-resolution measurements which are time consuming. In addition, storage and retrieval of rolls of ultraviolet-sensitive chart paper becomes cumbersome (especially from GC/MS runs).

It is therefore clear that computerized data acquisition and handling is potentially a valuable time-saving device in mass spectrometry. A fully automatic on-line mass spectrometer–computer system can produce, manipulate, store, and retrieve data much faster than the manually operated mass spectrometer system and can also effect comparisons with stored libraries of spectra in a matter of seconds.

The function of the computer in mass spectrometry is seen to best advantage where large amounts of data are collected and require interpretation. For example, computerized acquisition and processing of GC/MS data enables mixture constituents to be rapidly determined. The extensive applications of this combination (e.g., in perfumery, environmental studies, and drug analyses) attest to its usefulness.

The importance of computers in mass spectrometry requires no further emphasis, but the system should not be used merely for the acquisition of vast amounts of data without efficient interpretation. There is also a danger in extensive manipulations of data by a complicated, unseen computer program (e.g., some of the routines for cleaning up GC/MS data). As far as the hardware is concerned, the reliability of such complex equipment can be a problem but the technology continues to improve.

The choice of the type of mass spectrometer–computer system adopted is governed not only by the basic research requirements of a laboratory but also by economic factors (see Chap. 12 for factors to be considered in such a choice). The development of a variety of systems has therefore occurred, operating at various speeds, capable of widely different kinds of data manipulations and incorporating a variety of output devices and storage capabilities. The computer may also control aspects of the mass spectrometer (and GC) operation, including automatic sampling, injection, and scanning. Whichever system is chosen, other peripherals or software may frequently be added later, since most data systems are additive to a certain extent.

It is not the purpose of this chapter to give details of hardware, software, and mathematical operations used in mass spectrometer–computer systems. Rather, the basic principles of operation of the systems are outlined and a more detailed account may be found elsewhere.[1] In view of the large number of systems in operation the discussion must necessarily be abbreviated. Definitions of some of the terms applicable to this chapter are given below:

Algorithm	a sequence of computational steps for solution of a particular problem.
Analog / Digital	an analog variable is continuously changing whereas a digital variable is quantized.
Capacity	(of core or disc) number of computer words storable, usually quoted in units of 1024 (K).
Computer system	an assembly of interconnected machines containing input devices, storage devices, a central processing unit, and output devices.
Core	a device in the computer for magnetic storage of data.
Disk	a random-access device used for magnetic recording and storage of programs and data.
Hard copy	usually a paper copy of the image viewed on a visual display unit (VDU).
Hardware	the actual computer and equipment.
On-line	used to describe the direct connection of peripheral equipment (e.g., the mass spectrometer) to the computer.
Real-time	in real-time operation, the computer processes data as they arise and keeps in time with the arrival of new data.
Software	programming schemes used by the computer.

9-2 DATA ACQUISITION AND PROCESSING

A. Acquisition

There is a variety of operational mass spectrometer–computer systems suited to individual laboratory requirements and circumstances. Nevertheless, the basic principles of the computer-linked system are relatively simple. The output from an electrical-recording mass spectrometer is a voltage (typically a maximum of 10 V) which varies continuously with time as the spectrum is scanned. Figure 9-1 illustrates the voltage output over a single peak in a mass spectrum. The continuous electrical output (i.e., analog signal) from the detector of the mass spectrometer is sampled at precise intervals and converted into pulsed (i.e., digital) form by an analog-to-digital (A/D) converter. These digitized signals, consisting of an amplitude and a time (or count number), are then passed to the computer for processing.

A variety of sampling (digitization) rates may be chosen depending on the scan rate and the resolution required. Commercial data systems rarely employ digitization rates greater than 100 kHz (100,000 samples per second) and such high rates are only required for fast scans under high-resolution conditions. For low-resolution scans, A/D rates of 1–5 kHz are usually sufficient for scan rates of 3 s per decade in mass ($m \rightarrow 0.1\ m$), giving 10–20 digital samples per peak.

The next step is the thresholding of the data or rejection of digital samples having an amplitude below a certain threshold limit, so that only the digitized data for peaks remain (see Fig. 9-1). The threshold is sometimes lowered at high mass to allow recognition of weak but important high-mass peaks. Another signal-processing operation is the rejection of noise spikes (e.g., the discarding of peaks having fewer than five digital samples). Both these operations are usually performed by a specialized interface under software control.

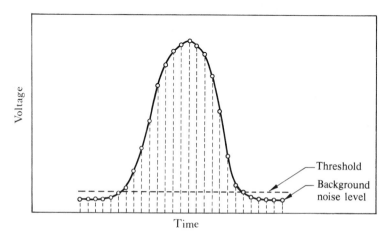

Figure 9-1

After thresholding, the data are ready to be reduced in real time to just two quantities for each peak, namely, the centroid time and peak area. Various formulas are available for calculating the centroid of the peak which theoretically has a gaussian shape. However, peaks of low intensity may contain only a few ions at maximum scan speeds. This results in irregular peak shapes so that the centroid is difficult to determine.

B. Calibration

Accurate masses of all peaks are then calculated by relating the scan time of unknown peaks to those of a reference compound whose spectrum and exact masses of peaks are known. The mass/time relationship (i.e., calibration curve) is constructed for low-resolution spectra by first obtaining a spectrum of the reference compound under reproducible conditions. The reference generally used is perfluorokerosene (PFK), which produces a series of peaks of accurately known mass at intervals of 12 and 19 mass units apart to above mass 800 (see Table 1-1). Another reference compound, perfluorotributylamine [$N(C_4F_9)_3$], gives a spectrum extending to just above mass 600 (see Table 1-4). A series of perfluoro-trialkyl triazines have been found useful for mass determinations up to 1500 (see Tables 1-2 and 1-3). For example, perfluorotriheptyltriazine (**1**) yields a molecular ion at m/z 1185 and fragment ions at masses 1166, 928, 866, 771, and 471 among others. The trinonyl homolog gives a spectrum extending downward from a molecular ion at m/z 1485. However, the reference peaks in these triazine spectra are too widely spaced in parts of the spectrum and this can be a drawback in automated data collection. To circumvent this problem, PFK/triazine mixtures have been employed for calibration up to mass 1500. For masses greater than 1500, the vacuum oil Fomblin (**2**) has been successfully employed for calibration, giving peaks at 16 and 18 mass unit intervals up to mass 2500.

1   **2**

The computer usually constructs the mass/time calibration curve by recognizing a few prominent peaks in the reference spectrum, fitting a partial curve from these data, and extrapolating stepwise through the rest of the spectrum, constructing the calibration curve *en route*. A different type of polynomial is employed depending on whether the scan law is basically linear (quadrupole) or exponential (magnetic sector). When using PFK, it is common for the computer first to identify masses 28, 32 (from air), and 69 (base peak from PFK) and then to attempt the identification of other masses (81, 100, 119, etc.) within a specified mass window. When more than a certain number (say five) of successive reference peaks are missed, the calibration is terminated. The operator would normally choose the best one of several scans (i.e., the one which calibrates to the highest mass with the smallest number of missed reference peaks *en route*). The upper limit of calibration for a given reference compound varies somewhat with the type of instrument and also tends to drop a little as the source and slits become dirty.

When running a sample spectrum using this created calibration file it is imperative to employ the same scan-start parameter and scan rate. The calibration can hold for a period of a few hours or several days, depending on the instrument, and it is common to check the calibration once or twice a day. In some modern data systems more reliable calibration is attained by computer *control* of the scan.

C. Interscan Reporting

A summary of the acquisition process is shown in Fig. 9-2. Scans are taken repetitively on the sample and in all but the simplest mass spectrometer data systems information about each spectrum is conveyed to the user between scans in the form of an *interscan report*. This information is presented on a teletype, line-printer, or visual display unit (VDU) and may be in numeric form. It includes such

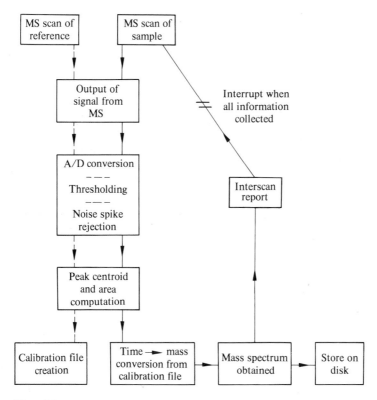

Figure 9-2

Table 9-1

Scan number	Base peak	Most intense peaks: intensity (%)			Highest mass peaks: intensity (%)		Number of peaks	Total ion current
1	57	43:73	71:68	85:53	141:2	128:1	65	210
2	57	43:72	71:70	85:49	198:1	197:5	75	245
3	57	43:75	77:70	71:68	198:5	197:40	115	510
4	77	180:90	197:67	57:60	211:1	198:15	190	1324
5	77	180:88	197:69	51:61	211:1	198:14	210	1556

items as the masses of the most intense peaks, highest mass peaks, and the total ion current. Table 9-1 shows the interscan reports for the first five scans on a probe sample of benzophenone oxime (**3**). Scans 1 and 2 show the spectra of residual hydrocarbon material which are superceded by the spectra of the sample as the probe warms up and the total ion current increases. Scans 4 and 5 indicate also that there is a small amount of homologous material present (at mass 211).

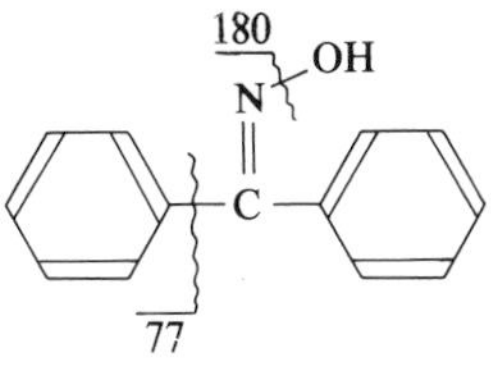

3, $M^{\dagger} = 197$

The speed and viewing clarity of a VDU are obvious advantages, and where one is available the output can take the form of a bar-chart mass spectrum. Other calculations may be performed in real time and the results presented in the interscan report (e.g., peak shapes, GC profiles), but it should be borne in mind that an increased amount of interscan information necessitates a longer interscan delay to perform the calculations. An interscan delay of about 1 s is usually sufficient to perform necessary calculations and in the case of magnetic instruments this period allows the magnet current to return to its initial value.

The sequence of scans is then continued until all necessary information is collected (e.g., at the end of a GC/MS run or when the direct insertion probe has reached a sufficiently high temperature). The spectra are then stored by the data system (on disk if available) for manipulation and output as required.

D. Data Manipulations

When a series of low-resolution mass spectra has been stored on disk (or magnetic tape) the computer may perform several operations on the data before outputting the chosen spectrum (or spectra) in graphical or alphanumeric form.

Some of these operations are peculiar to GC/MS data and will be considered below in Sec. 9-3.

The commonest data manipulations for probe samples are:

1. spectral averaging and
2. background subtraction.

Both are simple mathematical computations. Spectral averaging is sometimes necessary when the total sample pressure (as monitored by the total ion current) fluctuates during a scan, so that some peaks might be accentuated relative to others. Averaging of spectra from several adjacent scans improves the data. Spectral averaging also enhances the quality of field desorption spectra, in which statistical fluctuations are large. Background subtraction is self-explanatory and is most widely used to improve GC/MS or LC/MS data. However, there are occasions (e.g., the presence of background spectra from a contaminated source) where the quality of probe spectra is improved by this operation.

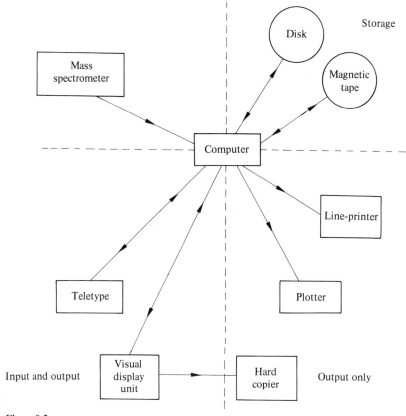

Figure 9-3

E. Output of Spectra

Figure 9-3 illustrates the various input, output, and storage devices that may be available with a mass spectrometer data system. For graphical output (i.e., a bar chart) there are basically two alternatives:

1. A "hard copy" may be obtained of the VDU display.
2. The spectrum may be plotted on a mechanical plotter.

The second alternative is usually slower but cleaner. However, some hard-copiers produce spectra of good quality.

If alphanumeric output is required, either alternative (1) above is employed or data may be printed on a teletype or on a line-printer. It should be noted at this juncture that packages are commonly available which allow the user access to the acquired data via the high-level FORTRAN or BASIC languages. The user can thus output data in a particular format with individual headings and data manipulations.

The above discussions have outlined some of the processes that occur in conversion, by the data system, of the signal leaving the mass spectrometer into a readable and acceptable low-resolution mass spectrum. Many sophisticated routines are available with modern data systems; some of these are illustrated in Table 9-2. In particular, systems are widely employed which acquire data simultaneously from two or more mass spectrometers. Another useful feature is the "spooling" of data in which output sent to a hard-copying device may be stored in

Table 9-2 Some features available with mass spectrometry data systems

	Feature	Purpose
1	Background subtraction	Cleaner spectra
2	Library searching	Identification of unknowns
3	High-resolution software	Element map production
4	Computer control of scan law	More reliable mass determination
5	Multiple-ion monitoring	Sensitive detection
6	High dynamic range	Detection of important weak peaks
7	FORTRAN or BASIC access to data	User-oriented output format
8	GC data reconstruction	Enhanced GC resolution
9	Simultaneous acquisition from two or more mass spectrometers	Full utilization of MS capability
10	Real-time diagnostics	Corrects mass spectrometer drift
11	Spectral averaging	Reduces statistical variations
12	Color VDU	Output clarity
13	High A/D conversion rate	Aids fast-scanning and high-resolution scans

a disk file and "emptied" in the background while the console terminal is used for further foreground operations. In concluding this section, it should be emphasized that many of the items, both hardware and software, in data systems are additive. Therefore, it is possible to add some of the features shown in Table 9-2 when requirements and financial circumstances allow.

F. High Resolution

Before data systems became available in mass spectrometry, the most frequently used method for accurate mass measurements was the peak-matching technique (see Sec. 1-5). This manual method is accurate but slow, and measurements are performed on only one mass peak at a time. It is therefore inappropriate for GC/MS spectra. However, there are now several routines available with data systems which determine accurate masses of all ions in the spectrum directly from scan data. In this section the limitations and advantages of these alternatives will be briefly considered.

A common routine for high-resolution analysis from double-focusing magnetic instruments necessitates the use of an *internal* standard. Calibration is performed in the usual way but unlike the low-resolution routines described above the sample is then introduced simultaneously into the ion source with the reference compound (for example, PFK) under conditions of high resolving power (5000 or higher). The output from the mass spectrometer therefore incorpo ates overlapping spectra of the sample and reference. The data system accurately identifies unknown masses by recognizing PFK peaks and interpolating between them. In this way, deviations and drifts in scan parameters, which could compromise results if an external standard were used, do not lead to excessive loss of accuracy.

It is possible to obtain precisions between 5 to 10 ppm in mass measurement from single scans by data system methods. Although such results are inferior to those obtainable by peak matching, they are sufficient in most cases to determine elemental compositions (see also Sec. 1-5). When the accurate masses of peaks in the spectrum have been determined, the computer calculates (in real time if the interscan period allows) the elemental composition(s) appropriate to each mass. Since all masses in the spectrum are being considered, it is important to limit the

Table 9-3

Integral mass	Composition				Observed mass	Error ppm	Base peak, %
	C-12	C-13	H	O-16			
74	3	0	6	2	74.0365	−3.7	100.0
87	4	0	7	2	87.0454	9.2	72.3
149					149.0236		10.4
267	18	0	35	1	267.2704	6.1	5.4
298	19	0	38	2	298.2876	1.5	15.0
299	18	1	38	2	299.2890	−5.1	4.6

numbers of elements considered by the computer, otherwise the calculation could take minutes rather than seconds. The output from such a computation is known as an "element map" and is illustrated by part of the output for the high-resolution mass spectrum of methyl stearate in Table 9-3. Only the elements C, H, and O were considered and the maximum number of oxygen atoms was limited to two. All the peaks given in Table 9-3 were correctly identified, with the exception of m/z 149. This corresponds to the well-known fragment ($C_8H_5O_3{}^+$) from phthalate ester plasticizers (see Table 6-5).

Production of element maps as given in Table 9-3 is routine on many high-resolution mass spectrometry–computer combinations, but a few limitations should be borne in mind:

1. PFK and sample pressures should not differ greatly and this situation is difficult to achieve in GC/MS runs.
2. Accuracy of mass measurement tends to depreciate below mass 100, since adjacent PFK peaks are not sufficiently close for accurate interpolation in this region of the scan.
3. Increased resolving power leads to a sensitivity loss, because of the partial closure of resolving slits. For example, an increase in resolving power from 1000 to 10,000 could require a tenfold increase in the minimum amount of sample required.

The double-beam mass spectrometer possesses unique advantages in the computation of accurate masses. Spectra of the sample and reference compounds are obtained in separate beams so that accurate masses in the sample spectrum are computed by comparison of centroid times between the two beams. Advantages of this mode of operation are:

1. The maintenance of similar source pressures for sample and reference is not a problem.
2. Overlap of mass spectral peaks does not present a difficulty if the peaks are found in separate beams. Consequently, adequate results may be obtained at lower resolving power (for example, 1500) than would be necessary with single-beam double-focusing instruments.
3. Accurate mass results can be obtained in the CI and FD modes by generating an EI reference spectrum in the other beam.

The accuracy achievable by double-beam accurate mass measurement is 10–20 ppm, and the merits of this mode, as outlined above, render double-beam software a powerful asset to a data system.

The precise control which computers are now able to exercise over the scan law of a mass spectrometer has led to the development of software packages that enable accurate masses to be obtained even from instruments of low resolving power (both magnetic and quadrupole). For example, if PFK is used as an external calibrant, as outlined previously, the scan law can be reliably maintained under computer control for at least several hours. As an additional safeguard, a second reference which has a few widely spaced peaks (for example, C_2I_4) may be introduced simultaneously with the sample and its mass spectrum used to correct any deviations in the scan law. A mass accuracy of better than 20 ppm is possible using this technique, but overlapping peaks are not resolved.

9.3 GAS CHROMATOGRAPHY–MASS SPECTROMETRY–COMPUTER SYSTEMS

Before the advent of data systems in mass spectrometry, identification of mixture components by a GC/MS system was often a time-consuming task. Adverse factors which hindered structural elucidation included:

1. A complex GC.
2. Small quantities of sample on a high background (from column bleed, etc.).
3. Overlapping GC peaks.
4. Vast amounts of chart paper which often required manual mass counting.

Section 8-2 has shown the current "state of the art" in GC/MS capabilities. This section demonstrates various ways in which computerized data systems may assist the fast extraction of structural information from a GC/MS injection. The techniques discussed here are also generally applicable to LC/MS analysis. Library searching will be considered separately in Sec. 9-4.

Mass spectra are normally taken continually during the entire chromatogram. Each scan period plus the interscan delay may be less than 1 s or as much as 10 s depending on the GC column and mass spectrometer resolutions employed. This leads to a total of as few as 50 scans and up to as many as several thousands. Where a data system has a limited storage capacity not all scans are stored.

A gas chromatogram is usually reproduced by the computer in the form of a total ion current (TIC) versus a scan-number trace (see Fig. 9-4a). The TIC is evaluated between each scan either by the hardware directly from the TIC monitor or by software summation of individual ion intensities.

The simplest manipulation on the acquired data is a subtraction of one selected background spectrum from the spectrum of an eluted peak. This operation corrects for column bleed and other background material. For example, absolute subtraction of scan 40 from scan 56 shown in Fig. 9-4a would produce a relatively clean spectrum of the compound eluted in peak *D*. Where background varies in different parts of the chromatogram (e.g., during a temperature-programmed GC/MS run) it is necessary to multiply the background spectrum by an appropriate scaling factor before subtraction.

The discussion in Sec. 8-2C has shown that single and multiple ion monitoring can be used for highly sensitive qualitative and quantitative detection of organic compounds. It is now commonplace for the computer to control these processes. With both magnetic and quadrupole mass spectrometers as many as 20 selected ions can be monitored in cycle times of 1 s or less. Where double-focusing magnetic instruments are employed it is possible to locate each mass within an accuracy of 20 ppm, thereby eliminating interfering ions. If quantitative measurements are

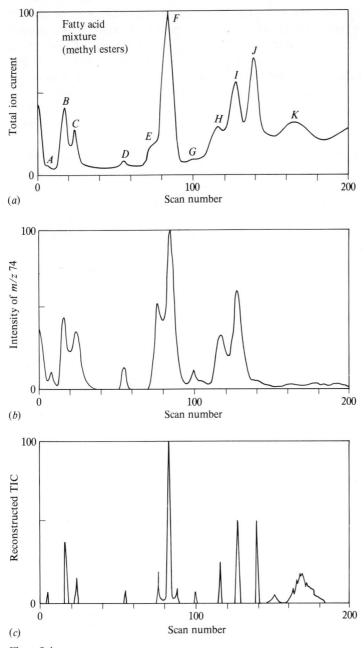

Figure 9-4

required (see Sec. 8-5) then the computation of relative peak heights and areas improves speed and accuracy.

The above techniques of multiple ion detection have some limitations for structural identification since major ions to be monitored are selected *before* a GC/MS run. However, non-real-time "mass chromatograms" may be readily constructed from stored data and used for structural analysis of mixture components. For example, a plot of m/z 74 (indicative of a —$CH_2CO_2CH_3$ group plus a γ-hydrogen to the carbonyl group), as shown in Fig. 9-4b, yields structural information on the fatty acid ester mixture whose total chromatogram is shown in Fig. 9-4a. Peaks J and K clearly do not incorporate this structural feature.

Most of the GC peaks shown in Fig. 9-4a are not completely resolved from one another and to identify unresolved peaks a simple reconstruction routine[2] may be applied to the data (in real time if required). The routine utilizes the fact that each GC component generates a number of ions in its mass spectrum which are unique. As shown in Fig. 9-4b, for m/z 74 the current for an ion of a particular mass may rise and fall with the elution of one GC component and not of another. In the reconstruction routine, when the ion current for a given mass maximizes, the scan number is "flagged" (or "noted"). For example, scans 8, 15, 24, 55, 76, 86, 99, 116, and 129 are flagged for m/z 74 in Fig. 9-4b. This operation is performed for all masses, the individual flagged ion intensities are stored, and a reconstructed TIC trace is generated by summing these flagged ion currents. The result is an enhanced GC trace (Fig. 9-4c) in which even small shoulders are resolved as separate peaks.

A reconstructed chromatogram may be generated in real time and followed on a VDU. However, the reconstructed mass spectra generated by this simple routine are often of limited value and more sophisticated software is required for production of "pure" spectra from GC traces. Programs have been described[3] which compute mass spectra in which column-bleed background and interference from neighboring elutants have been removed. Components that elute within less than two scan index numbers of each other can be detected and their mass spectra can be well resolved. Analysis of complex chromatograms by such programs usually takes several minutes.

9-4 LIBRARY SEARCHING

Having obtained a mass spectrum with the aid of a data system, the user might be confronted with the two questions:

1. Has the spectrum been seen before in this laboratory?
2. Has it previously been seen elsewhere?

Recourse to computer-stored mass spectral libraries will not always provide the answer (especially to the second question, since most known organic compounds do not have computer-stored mass spectra). However, in many cases, sensible searching of an appropriate library will provide rapid identification of an unknown. In this section the principles involved in the comparison between mass spectra will be discussed, followed by a presentation of some of the methods employed to search mass spectral libraries with a view to fast and accurate identification.

A. Comparison of Mass Spectra

The choice between various computational routines used for comparing mass spectra usually reduces to a compromise between speed and reliability. The most time-consuming calculations involve comparisons of *whole* spectra (including intensities as well as masses) whereas the fastest comparisons incorporate algorithms which consider only a few important peaks and neglect intensities. Accuracy of identification is sometimes lost in making the comparison too simple and in practice the methods adopted usually lie between the two extremes mentioned above.

In any method for comparison of mass spectra it is essential that the spectra are all normalized by the same method. Various normalization procedures have been used and the simplest of these is to make the sum of all the k observed peak heights P_n equal to unity:

$$\sum_{n=1}^{k} P_n = 1 \qquad (9\text{-}1)$$

Intensities are then compared between the two compounds over all n observed m/z values in both spectra. A suitable similarity index (SI) may be defined as follows:

$$\text{SI} = \left[1 - \sum_{n=1}^{k} \left| \frac{P_n(\text{ref}) - P_n(\text{unknown})}{2} \right| \right] \times 100 \qquad (9\text{-}2)$$

where $\text{SI} = 100$ for a perfect correlation and $\text{SI} = 0$ for no similarity. For a variety of reasons (e.g., measurement of spectra on different mass spectrometers, with different source residence times, operating temperatures, etc.), SI for identity will be close, but not equal, to 100.

In practice, calculation of a similarity index is usually not as simple as portrayed above. It is common to modify Eq. (9-1) by normalizing a *function* of P_n to unity, e.g.,

$$\sum_{n=1}^{k} P_n^2 = 1 \qquad (9\text{-}3)$$

B. Types of Search

It is wasteful of computer time to compare *all* ions in the unknown and library spectra, so abbreviated comparisons have been devised. For example, a search method is frequently employed which considers only the eight most intense peaks in the unknown and reference spectra. The number 8 is arbitrary and 4, 6, 12, etc., peaks have also been used. Another search technique used, particularly when the

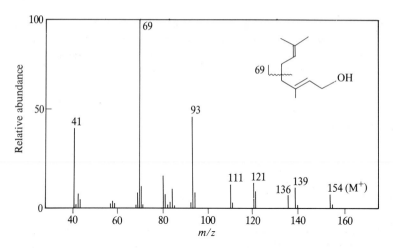

Figure 9-5

mass range is wide, normalizes and compares the most intense peaks (usually 1 or 2) within a series of narrow mass ranges, for example, 21–34, 35–48, 49–62.

Two of the most common search procedures (forward and reverse searches) employ opposing philosophies. In the *forward* search mode, a selected unknown spectrum is compared in turn with a series of library spectra. If the unknown spectrum is reasonably pure and/or a large library is to be searched, then the forward mode would be a likely choice.

An example of the output from a forward library search is shown in Table 9-4. The spectrum of geraniol, obtained from scan 64 of a GC/MS analysis of a perfume mixture (Fig. 9-5), is compared against a library of known perfume components (see also Sec. 10-3 for a related analysis). A minimum similarity index of 80 percent is applied as a criterion for possible identity. The computer search identifies the most likely component as geraniol itself (similarity index 96) with its geometric isomer nerol being only slightly less probable. Other monoterpene isomers such as linalool (acyclic tertiary alcohol) and borneol (cyclic) have quite different spectra and are not therefore identified by the search. Geranyl acetate (SI = 81) is not the correct compound, since its molecular weight is 196 and its spectrum contains an abundant m/z 43 ion due to the presence of the acetyl group.

Table 9-4 Scan 64, three fits

SI, %	Base	Molecular weight	Reference	C	H	O	Name
96	69	154	T12	10	18	1	Geraniol
93	69	154	T14	10	18	1	Nerol
81	69	196	T45	12	20	2	Geranyl acetate

Table 9-5 Approximate forward and reverse library search times

Library size (number of spectra)	Forward search	Reverse search
200	0.5 s	5 s
2,000	5 s	50 s
20,000	50 s	500 s

The report shown is in an abbreviated form but the reference number included, for example, T12 for geraniol, provides an identifier for the extraction of extended information (e.g., the full library spectrum) from the computer disk.

The *reverse* mode of library searching follows the strategy of selecting a library spectrum and searching for a fit with a series of unknowns. The process is repeated for a series of library spectra. Each comparison is effected by selecting only those masses in the unknown spectra which occur in the library spectrum. This allows high background peaks in the unknown to be rejected and appropriate normalization permits a reliable comparison. The reverse search is useful where the operator has some idea of the structure and where a high background is likely to occur (e.g., where column bleed is high relative to an eluted peak in a GC/MS run or where there is a "memory" effect in an LC/MS run).

Reverse searching with normalization is usually slower than forward searching for libraries of similar size; this is illustrated in Table 9-5. Where small libraries are employed the speed of the forward search enables identification to be presented in the interscan report. Such methods have been found particularly useful for the rapid analysis of drugs of abuse from "overdose" patients.

Table 9-6 A sequential peak search of a mass spectral library

Ions specified	Intensity limits, %	Number of compounds found	Examined products
m/z 112	60–100	225	—
$+ m/z$ 154	5–40	19	—
$+ m/z$ 139	10–50	5	3 isomers of menthone plus 2 cyclopentanone derivatives
$+ m/z$ 69	40–100	3	3 isomers of menthone,

There are numerous variations in the procedures used for library searching, many of them tailored to the requirements of particular laboratories. A further technique, worthy of special mention because it is simple and effective, involves the specification of only one mass spectral peak at a time from the unknown spectrum. This procedure necessitates more user participation and is widely used in searches of large libraries. An example is shown in Table 9-6 of a sequential peak search on an authentic sample of menthone, whose mass spectrum contains the four most intense peaks at m/z values 69, 112, 139, and 154 of respective intensities 70, 100, 40, and 30 percent.

The library is first searched for all spectra having a peak at m/z 112 between 60 and 100 per cent relative abundance. This constraint is fitted by 225 compounds, and elimination of some of these possibilities is achieved by specification of further peaks until a sufficiently low number (say less than 10) of fits is found. The names of the compounds are then examined and more peaks are specified if necessary. In the example quoted in Table 9-6, specification of four masses was necessary to identify the unknown as menthone. It should be noted that full spectra of identified compounds can be examined at any stage, but it is only convenient to do this when the number of compounds is small. Generally, identification is achieved more quickly if higher masses of average or higher intensities are specified first.

If the time taken for a mass spectral search is considered too long for a user's requirements it may be advantageous to impose "filters" to restrict the spectra searched. There are many cases where the search might be confined to compounds of a particular molecular weight or class, or containing certain elements. In addition, other characteristic physical parameters, such as GC retention times, may be available if the user's work incorporates a well-defined instrumental system. Values may be assigned to all the above-mentioned parameters and used to filter the spectra. For example, in a spectral search of insecticides containing chlorine and aromatic rings three filter values could be imposed if the library permits them.

In conclusion, there are basically three options available for the user of mass spectral libraries:

1. A dedicated library may be constructed for operation with the user's own data system. This is advantageous where a small library of reliable spectra is sufficient.
2. A commercial library (20,000 or more spectra) may be purchased on disk and operated as above.
3. A commercial library may be searched via a telephone link.

C. Structural Recognition Programs

The above-mentioned library search programs have based their success or failure purely on a direct comparison of mass spectra (i.e., comparison of one "fingerprint" with another) without any recourse to recognition of structural features from examination of the spectra.

Library search routines have been developed which rely partly on structural-recognition software. For example, the self-training interpretive and retrieval system (STIRS)[4] is designed to extract structural information from unknown mass spectra, even when the authentic spectrum is not present in the library. STIRS directly utilizes the reference spectra and the known substructures present in the corresponding compounds to suggest structural units present in the unknown. For example, a number of classes of mass spectral data known to have high structural significance (e.g., characteristic ions, masses of neutrals lost) are identified. For each class the computer matches the data of the unknown mass spectrum against the corresponding data from all reference spectra. Reference compounds with high "match factors" in each class are then examined for frequently occurring substructures. Despite having no prior spectra-structure correlation, STIRS is able to provide structural information about complete unknowns.

The ultimate computer program for identification of unknown compounds from their mass spectra would be one which determines either structural features or the complete structure from the mass spectrum without recourse to library comparisons. The development of the DENDRAL family of programs has been underway for over a decade with this aim in mind.[5] For some compound classes (e.g., aliphatic ketones), the success of DENDRAL has been impressive. However, mass spectral computer programs currently seem to be some way from invariably exceeding the success of a skilled human interpreter in structural elucidation (partial or complete) from mass spectra of unknown compounds.

CONGEN, one of the programs of the DENDRAL package, has been shown to be a valuable aid on occasions to the organic spectroscopist.[6] This sophisticated program utilizes data from *all* the spectroscopic techniques (MS, NMR, infrared, ultraviolet, etc.) to generate all feasible two-dimensional structures (acylic and cyclic) from constraints imposed by the spectroscopic information. The program typically runs for periods from several minutes to several hours and generates structures that can be displayed spatially on a standard teletype or on a VDU. The program requires the following information:

1. The molecular formula. This is usually obtained from a high-resolution mass spectral measurement.
2. Detailed structural features. For compound **4** below this would entail:
   (a) the presence of the carbonyl and NH groups from the infrared spectrum and Br from the mass spectrum,
   (b) the presence of the *p*-disubstituted benzene function and the C_2H_5 group from the NMR spectrum,
   (c) any information connected with the synthesis of the compound.

$$Br-\langle\!\!\!\!\!\bigcirc\!\!\!\!\!\rangle-NH\cdot CO\cdot CH_2\,CH_3$$

4

The main advantage of CONGEN is that the prejudice of the spectroscopist for a particular structure is overridden. For structures as simple as **4** this is not usually an important factor, but for complicated structures CONGEN reveals structural possibilities not considered by the interpreter.

REFERENCES

1. J. R. Chapman: "Computers in Mass Spectrometry," Academic Press, New York, 1978.
2. J. E. Biller and K. Biemann: *Anal. Lett.*, vol. 7, p. 515, 1974.
3. R. G. Dromey, M. J. Stefik, T. C. Rindfleisch, and A. M. Duffield: *Anal. Chem.*, vol. 48, p. 1368, 1976.
4. K-S. Kwok, R. Venkataraghavan, and F. W. McLafferty: *J. Am. Chem. Soc.*, vol. 95, p. 4185, 1973; G. M. Pesyna and F. W. McLafferty: In F. C. Nachod (ed.), "Determination of Organic Structures by Physical Methods," vol. 6, Academic Press, New York, 1976.
5. J. Lederberg, G. L. Sutherland, B. G. Buchanan, E. A. Feigenbaum, A. V. Robertson, A. M. Duffield, and C. Djerassi: *J. Am. Chem. Soc.*, vol. 91, p. 2973, 1969.
6. R. E. Carhart, D. H. Smith, H. Brown, and C. Djerassi: *J. Am. Chem. Soc.*, vol. 97, p. 5755, 1975.

SPECIFIC ANALYTICAL APPLICATIONS

During recent years the numbers of activities which employ mass spectrometry as an analytical tool have continued to proliferate. This is because mass spectrometer systems are capable of the rapid identification of trace levels of compounds having widely differing mass and chemical structure (including mixtures). The large range of applications requires a variety of inlet systems, ion source types, mass analyzers, mass resolution, and mass ranges. Consequently, the selection of the most appropriate combination of mass spectrometric techniques for a particular application is often difficult and constantly under review (see Chap. 12). In this chapter, a few of the more important and interesting analytical topics have been selected for investigation. Comprehensive accounts may be found in leading review articles.[1,2]

10-1 BIOMEDICAL APPLICATIONS

The vast majority of biological and medical applications of mass spectrometry incorporate GC/MS instrumentation (see Sec. 8-2), employing both electron-impact and chemical ionization sources. LC/MS combinations (see Sec. 8-3) offer promise for the future. Computer techniques are also widely utilized (see Chap. 9).

Both urine and serum samples are commonly used for biomedical analyses by GC/MS but many tissue types and other body fluids have been examined by this technique. These include tissues such as brain, kidney, liver, lung, pancreas, and fat, together with salival, amniotic, seminal, synovial, cerebrospinal, and gastric fluids.

The most difficult aspect of the analysis is frequently the sample work-up and chemical derivatization for GC/MS. Separation procedures that have been widely used include solvent extraction, gel filtration, ion exchange, liquid chromatography, and membrane filtration. This battery of techniques enables the isolation of almost every class of metabolite, covering the whole range of molecular polarities. The method of derivatization must be chosen with care, the main considerations being the yield of the derivatization procedure, the stability and volatility of the derivatives, and their suitability for mass spectrometric analysis (see also Sec. 8-2B). It is not always possible to choose an appropriate derivative for GC/MS analysis and alternative methods must then be found.

A. Metabolism Studies

Determination of metabolites in the body by GC/MS methods has now become commonplace. In particular, GC/MS *profiling* has enjoyed considerable success in the diagnosis of certain metabolic diseases. The profiling approach aims to identify as many metabolites as possible in a given urine or serum sample. Profiles (i.e., complex GC traces) are compared for normal patients and those under study. The reduction or elevation of the levels of particular compounds or the detection of abnormal metabolites often provides clues about the nature of a disease. For example, it is possible to identify metabolic disorders in new-born infants by GC/MS profiling. Computerized data handling greatly speeds up the diagnosis of such disorders.

Profiling is normally carried out for selected compound classes, separated by specific work-up procedures. In particular, analyses of organic acids, steroids, and carbohydrates have proved to be of considerable diagnostic value for metabolic disorders. Notably, many of the diseases detected by GC/MS profiling techniques are organic acidemias or acidurias (i.e., an excessive amount of certain organic acids, respectively, in the blood or urine). For example, specific metabolites for isovaleric acidemia are isovaleric acid itself, isovalerylglycine, and 3-hydroxyisovaleric acid.

One drawback of the current GC/MS profiling approach is the inability to detect involatile compounds. However, the use of LC/MS techniques offers considerable promise in this respect (see Sec. 8-3) and profiling of certain other compound classes (e.g., amino acids and peptides) is currently performed successfully by thin-layer or ion-exchange chromatography.

The main attraction of GC/MS techniques in metabolic studies lies in the identification of metabolites at low levels. An example taken from the early days of GC/MS neatly illustrates the capabilities of the technique in the identification of the steroid 5α-pregnan-3α,20α,21-triol (1) in human pregnancy plasma.[3] Only 2 μg of the compound were isolated and no authentic sample was available, so the compound was identified by a few microchemical transformations combined with GC/MS.

Table 10-1

| Derivative | Relative retention times | | Conclusions about unknown |
	Unknown	$5\alpha,3\beta,20\beta$ isomer	
Parent triol TMS ether	2.06	2.74	
17-CO$_2$CH$_3$ 3-ketone	0.74	0.74	5α
17-CHO 3-TMS ether	0.56	0.70	3α
Acetonide 3-ketone	1.35	1.23	20α

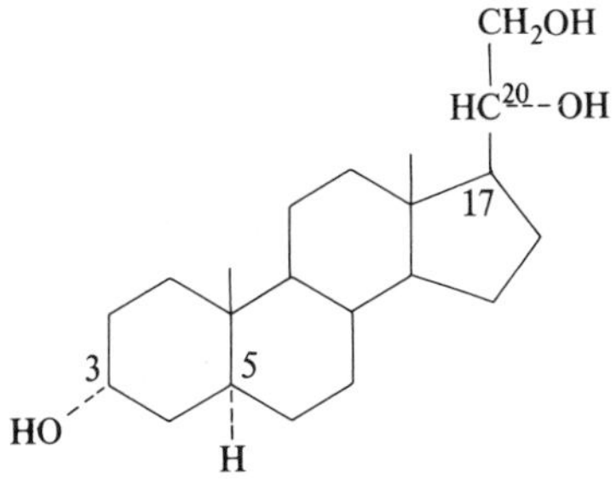

1

GC/MS analysis of the TMS ethers of steroids in the "monosulfate" plasma fraction identified a component having a molecular weight and mass spectrum corresponding to those of the authentic $3\beta,20\beta$ isomer of **1**, but possessing a different retention time. The unknown compound was therefore tentatively regarded as a pregnane-3,20,21-triol isomer. The configurations at the carbon atom 3 and 20 positions were established by comparing the retention times (SE-30 column) and mass spectra of some simple derivatives of the unknown and the authentic isomer. Table 10-1 lists the retention times of the derivatives, together with the conclusions reached about the unknown.

B. Pharmacology and Toxicology

Investigation of the mechanism of drug action in the body involves detailed determinations, both qualitative and quantitative, of the fate of the unchanged drug and detection of its metabolites. Pharmacologists have turned increasingly to mass spectrometry for these investigations in recent years. GC/MS has been extensively used for such studies and selected ion monitoring has become a commonplace detection procedure (see, for example, the determination of some chlorpromazine derivatives, Fig. 8-4). However, LC/MS is now being used for some investigations and the usefulness of thin-layer chromatography (TLC) to clean up metabolites should not be overlooked. Drugs have even been identified by their collisional activation spectra (see Fig. 8-9).

The techniques of quantitative mass spectrometry, which were presented in

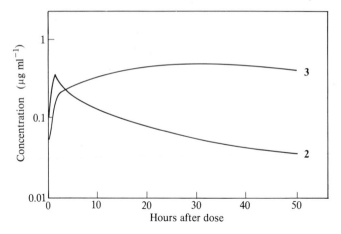

Figure 10-1

Sec. 8-5, are commonly used in pharmacology. One particularly important area of application is pharmacokinetics, where the rates of metabolism of drugs and their pharmaceutical preparations are determined.

An example of a pharmacokinetic study is shown in Fig. 10-1. The concentrations of Mephenytoin (**2**) and its demethylated metabolite Nirvanol (**3**), both of which are anticonvulsant drugs used, for example, in the treatment of epileptic patients, were measured in human plasma. A selected ion-monitoring technique was employed[4] and the assays were carried out at regular time intervals following single oral doses of Mephenytoin. Figure 10-1 shows the curves obtained for a 50 mg dose.

2, R = Me, Mephenytoin
3, R = H, Nirvanol

Mephenytoin was rapidly absorbed and then began to disappear from the plasma with a half-life of 12 h. Its metabolite, Nirvanol, slowly accumulated, reached a maximum level after 24 h, and then gradually disappeared from the plasma. It therefore appears that Mephenytoin is responsible for the early anticonvulsant activity whereas the accumulating concentration of Nirvanol maintains the activity for longer periods.

For quantitative measurements by mass spectrometry a procedure (including extraction methods, derivatization techniques, and GC/MS conditions) should be established separately for each sample type to be assayed (see Sec. 8-5). In the above example, both electron-impact ionization and chemical ionization were employed. The use of two ionization methods increases the reliability of the results.

Sample extracts were ethylated prior to GC/MS analysis and alkylated homologs were used as internal standards. The assay method adopted permitted levels of quantitation for **2** and **3**, respectively, of 10 and $50 \, \text{ng ml}^{-1}$. Measurements were repeated and found to be reproducible to within 8 percent.

The examples shown in Figs. 8-4 and 10-1 illustrate the applications of the techniques of mass spectrometry to the development of beneficial drugs. Such methods may also be used to analyze drugs of abuse. For example, efficient gas chromatography–mass spectrometry–computer systems are available in some hospitals for the rapid analysis of plasma from drug-overdose patients. Since a comprehensive list of these drugs is not likely to exceed 1000 compounds, such an analysis lends itself to rapid identification of the mass spectra via a computerized library search (see Sec. 9-4). Another related area for application of gas chromatography–mass spectrometry–computer techniques is the detection of drugs taken to enhance athletic performance in competition. The necessity for these analyses has arisen in human athletic activities and also in horse and greyhound racing. One major problem is the constant development of new "performance" drugs which sometimes leaves the drug analyst one step behind.

C. Protein Sequencing

Various systematic methods have been developed for determining the sequence of amino acids in proteins by mass spectrometry. A procedure which utilizes acetylation and permethylation of the peptides was discussed fully in Sec. 6-5D. In this section a different method is presented which involves selective cleavage of proteins into small peptide fragments, followed by derivatization and analysis by GC/MS. Some chemical and biochemical reagents employed in this type of work for peptide bond cleavage are indicated as follows:

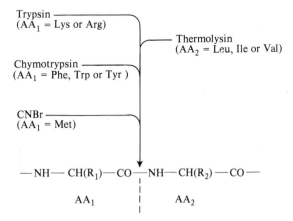

In a recent application of this strategy,[5] bacteriorhodopsin (248 amino acid residues) was initially split into smaller fragments by the enzyme chymotrypsin and by digestion with cyanogen bromide. The fragments were then digested further, either by another enzyme (e.g., thermolysin) or by partial acid hydrolysis. Mixtures of peptides so derived were methylated, trifluoroacetylated, and then reduced with either $LiAlD_4$ or B_2D_6. The resulting polyaminoalcohols were silylated and then injected onto the column of a GC/MS combination. Fragmentation of these compounds in the mass spectrometer (predominantly by α cleavage of amines, see Sec. 6-3C) permits sequence determination of the peptides. The use of deuteriated derivatives renders the spectra easier to interpret. The sequence of the full protein may then be identified from the small, structurally overlapping peptides. This complex problem is greatly assisted by the use of computerized data collection and interpretation. Even estimated retention indices of the constituent derivatized peptides on the GC column aid the identification. Furthermore, the complementary use of mass spectrometric and automated Edman sequencing is important.

An elegant method for identifying structurally overlapping peptide fragments from cyanogen bromide cleavage has evolved from the above general technique. Since cyanogen bromide cleaves the protein at an amide bond on the C-terminal side of methionine (Met), the sequence determination of the N-terminal fragment will give the sequence X-Met and identify fragments of m/z values characteristic of X-Met. Peptides not undergoing this specific cleavage possess the structure X-Met-Y. Mass chromatograms (see Secs. 8-2C and 9-3) are determined for the GC effluent, monitoring an ion characteristic of the known fragment X-Met. All the GC/MS data are stored in computer memory, scans exhibiting the selected ion(s) in high abundance can be retrieved, and, by examination of the full spectra, the nature of Y in X-Met-Y can be determined.

10-2 ENVIRONMENTAL ANALYSIS

The application of mass spectrometry to the analysis of organic pollutants in air, water, soil, and living organisms has inevitably developed rapidly in the past decade, owing to the capability of the technique in analyzing organic mixtures at trace levels. A vast amount of literature has accumulated in this field and review articles should be consulted for a more comprehensive account.[1,2,6]

The analysis of organic environmental pollutants may be broadly divided into two categories:

1. *Survey analysis.* This type of investigation aims at an extensive qualitative analysis of all organic compounds present in an environmental sample. Potential hazards may be identified by this approach. Analyses are often facilitated by computerized searching (see Sec. 9-4) of mass spectral libraries of potential pollutants.
2. *Analysis of individual pollutants.* This type of analysis helps to determine whether quality standards are being satisfied. Quantification (see Sec. 8-5),

either of individual compounds or compound classes which are known to possess hazardous properties, is carried out.

Both of these approaches nowadays increasingly employ mass spectrometry; some experimental techniques and results obtained for environmental pollution are considered below for atmospheric and aqueous samples. The section is concluded with a brief discussion of analysis of residues from soil and living organisms.

A. Atmospheric Pollution

The most widely used enrichment method for atmospheric pollutants having boiling points above about 60°C (e.g., less volatile than hexane) is "dynamic enrichment" on appropriate adsorption columns at ambient temperature.[6] The adsorbed compounds are then eluted for GC/MS analysis either by thermal elution (preferable) or solvent elution. Table 10-2 summarizes dynamic enrichment procedures using three different types of adsorbent column. Compounds are adsorbed onto the sampling column by drawing air through the apparatus. The pollutants are then preferably thermally eluted into the GC/MS combination by a purge gas flowing in the opposite sense to the sampling direction. This facilitates the desorption of higher boiling compounds.

Two compound classes extensively recognized as atmospheric pollutants are aromatic and halogenated hydrocarbons. Millions of tonnes of aromatic hydrocarbons are discharged annually into the atmosphere from industrial solvents, combustion, etc. The polynuclear hydrocarbons, many of them established carcinogens, are particularly amenable to analysis by single or multiple ion monitoring because of their abundant molecular ions following electron-impact ionization (see Sec. 6-1). However, it is difficult to distinguish between isomers.

Concern has been expressed that dissociation of halogenated hydrocarbons in the stratosphere may decrease the ozone concentration and adversely affect the level of ultraviolet radiation transmitted to ground level. Results have shown that these compounds are more persistent in the atmosphere than hydrocarbons. A variety of short-chain halogenated alkanes and alkenes have been identified by GC/MS techniques, particularly in metropolitan atmospheres. Their characteristic isotope patterns (see Fig. 6-1) are readily recognized and detection is possible at the parts per trillion level.

Table 10-2 Enrichment procedures for organic air volatiles by dynamic column adsorption

Adsorbent material	Method of elution into GC/MS
Porous polymer	Solvent or thermal elution
Graphitized carbon black	Solvent or thermal elution
Activated charcoal	Solvent elution only

B. Water Pollution

The analysis of organic water pollutants by mass spectrometry has become a specialized field of research, due to the impact that such pollution is having both on aquatic and terrestrial life. Despite the importance of this research, methods for detection of high-boiling water pollutants are not currently well developed and modern procedures cover mainly the analysis of volatile organic compounds. Among sampling and enrichment procedures employed are[6] (1) distillation and/or headspace analysis, (2) various continuous extraction procedures using organic solvents, (3) adsorption of pollutants on appropriate solid polymer materials, and (4) purging of organic volatiles from water using a carrier gas, with subsequent gas-solid adsorption. Where solvents are used in the above enrichment procedures, the samples are injected directly into the GC/MS combination. However, where solid-phase adsorption is used, thermal elution and transfer to the GC/MS by dry carrier-gas flow is an alternative analytical procedure.

Among the organic pesticides and herbicides which are detectable as water pollutants, halogenated aromatic compounds have proved to be particularly persistent. For example, the herbicide 2,4,6-trichlorophenyl-4'-nitrophenylether (CNP, **4**) was detected by GC/MS methods in Japanese rivers and agricultural drainage and was monitored over a period of several months.[7] An adsorption method was used for enrichment of the organic materials, which were then removed from the resinated column by solvent elution (diethyl ether). Positive identification of CNP at the nanogram level was facilitated by abundant molecular ions (following electron impact) and isotope patterns characteristic of three chlorine atoms. Maximum levels were detected in river water within one month of the beginning of rice-seedling transplantation and simultaneous herbicide application to the paddy fields.

4

C. Residue Analysis

Another application of mass spectrometry to environmental research is the analysis of residues from man-made biologically active molecules, such as (**4**) described above, which accumulate in soil or living tissue. Residue analysis has been carried out using mass spectrometric techniques on soil samples, plant extracts, and a variety of tissue, plasma, or urine samples from animals, birds, fish, insects, etc. Extraction procedures vary widely depending on the material and the contaminant. Individual publications should be consulted for experimental details.[6]

Nitrosamines, R-NH-NO, have attracted much attention because of their carcinogenic properties. Once again, GC/MS has proved to be a powerful method for analysis of nitrosamines in fish and meat products. High-resolution selected ion monitoring (see Sec. 8-2C) is particularly useful in this respect to identify the relevant N-containing low-mass ions.

The effects of oil spillage on the environment are investigated by mass spectrometric analysis not only of organic pollutants in water but also of residues which accumulate in living tissue. In these, as in other environmental studies, it is important to examine the fate of the pollutants at different time intervals after the spillage occurs in order to evaluate their persistence.

As indicated above for water-pollution studies, organochlorine pesticides and herbicides show alarmingly high residue levels. On account of the widespread use of these compounds, specific procedures have been developed for analysis of their residues. For example, the highly toxic 2,3,7,8-tetrachlorodibenzo-*p*-dioxin (TCDD) (5), a contaminant in some herbicide and fungicide preparations, has been analyzed by negative-ion chemical ionization mass spectrometry at the low nanogram level.[8] At a source pressure of 1 torr, O_2 produces high concentrations of O_2^- and O^- following electron impact. Chemical ionization of the TCDD molecule in this medium affords an abundant, structure-specific ion (6) at m/z 176, which is well out of the mass range of ions due to other contaminating chlorocarbons. Consequently, this ion may be used for sensitive and specific detection of TCDD by mass spectrometric methods.

Although GC/MS using electron impact or chemical ionization is the most frequently used technique for pesticide analysis, other mass spectrometric methods should not be overlooked. The use of LC/MS was illustrated in Fig. 8-7; alternative ionization methods such as field desorption, negative ionization, and atmospheric-pressure ionization have also been employed in this field.[9]

10-3 FLAVORS AND PERFUMES

The variety of "essential oils" (terpenoid compounds) from plants which impart the pleasant flavor or odor to a wide variety of perfumes, herbs, and beverages are amenable to analysis by GC/MS (usually without derivatization). Furthermore, because of their characteristic spectra and limited numbers, mixtures of essential oils are ideal for analysis by a library search using a dedicated computer. A comprehensive library for essential oils contains less than 1000 spectra and can be searched quickly (see also Table 9-4).

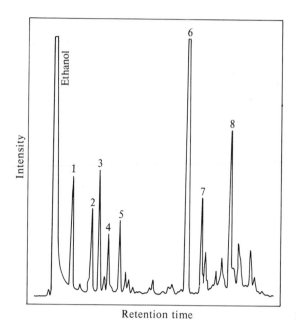

Figure 10-2

An example of the successful analysis of such a mixture by GC/MS is found in the identification of the flavor concentrates of gin.[10] Basically, all gins are made by distilling spirit in the presence of balanced botanical ingredients which impart the characteristic flavor. Juniper berries form the basis of any gin formulation and coriander seeds are frequently incorporated.

Figure 10-2 illustrates a GC trace (packed column) of the continuous liquid/liquid extraction product (see also Sec. 10-2B) from London Dry gin, using Freon 11 as solvent. Overlapping GC peaks may, of course, be resolved by alternative columns. Reconstruction of the extracts obtained using Freon 11 showed that they were identical organoleptically to the original samples.

Table 10-3 Some components of commercial gin identified by GC/MS

Peak number (Fig. 10-2)	Compounds
1	α-Thujene, α-pinene
2	Sabinene
3	Myrcene, 3-carene, α-phellandrene
4	Limonene
5	γ-Terpinene
6	Linalool plus unidentified sesquiterpene
7	Terpinen-4-ol
8	γ-Muurolene, citronellol

Over 50 terpenoid flavor components were identified in the gin, both from their mass spectra and from GC retention times, and some of the more abundant of these are shown in Table 10-3. A high proportion of the more volatile compounds are often juniper components, whereas constituents of coriander oil are found in greater frequency in the higher-boiling gin distillate.

Various gins produced from different base levels of juniper and coriander were "profiled" using the GC/MS approach (see Sec. 10-1A) and the profiles correlated with conclusions from human tasters. One criterion for a good-quality gin which emerged from the tests was that it should contain γ-muurolene (7) and 4-terpinenol (8) in approximately equal quantities. An excess of 4-terpinenol produces a poor-quality gin.

7   8

REFERENCES

1. R. A. W. Johnstone (ed.): "Specialist Periodical Reports," vol. 5, The Chemical Society, London, 1979.
2. A. L. Burlingame, T. A. Baillie, P. J. Derrick, and O. S. Chizov: *Anal. Chem.*, vol. 52, p. 214R, 1980.
3. J. Sjövall and K. Sjövall: *Steroids*, vol. 12, p. 359, 1968.
4. W. Yonekawa and H. J. Kupferberg: *J. Chromatog.*, vol. 163, p. 161, 1979.
5. H. G. Khorana, G. E. Gerber, W. C. Herlihy, C. P. Gray, R. J. Anderegg, K. Nihei, and K. Biemann: *Proc. Natl. Acad. Sci. U.S.A.*, vol. 76, p. 5046, 1979; see also H. Nau and K. Biemann: *Anal. Biochem.*, vol. 73, p. 139, 1976.
6. H. Knöppel: *Annali di Chimica*, vol. 67, p. 649, 1977; A. Alford: *Biomed. Mass Spectrom.*, vol. 5, p. 259, 1978.
7. M. Suzuki, Y. Yamato, and T. Akiyama: *Water Research*, vol. 12, p. 777, 1978.
8. D. F. Hunt, T. M. Harvey, and J. W. Russell: *Chem. Comm.*, p. 151, 1975.
9. G. Vander Velde and J. F. Ryan: *J. Chromatog. Sci.*, vol. 13, p. 322, 1975.
10. D. W. Clutton and M. B. Evans: *J. Chromatog.*, vol. 167, p. 409, 1978.

ELEVEN

ISOTOPIC LABELING

11-1 SYNTHESIS OF ISOTOPICALLY LABELED ORGANIC COMPOUNDS

In this section, only a brief account of the principles involved in synthesizing labeled compounds is presented; more extensive discussions are given elsewhere.[1] Experimental details are also available for a host of synthetic routes to labeled organic compounds.[2]

The main aim in most syntheses of labeled compounds is to incorporate the label as late as possible in the sequence of reactions. This strategy is often facilitated by the availability of ^{2}H-labeled solvents, designed for nuclear magnetic resonance (NMR) spectroscopy; for example, d_6-benzene, d_4-acetic acid, d_6-acetone, d_4-methanol, and d_8-toluene. These starting materials may be used directly or converted into other useful building blocks.

The most useful isotopes for mass spectrometric purposes are ^{2}H, ^{13}C, ^{15}N, and ^{18}O; incorporation of these isotopes into organic compounds may be accomplished in a variety of ways.

A. Incorporation of Deuterium

Deuterium is by far the most common isotope used in mass spectrometric studies, since it is easy to incorporate and because essentially isotopically pure (99.7 percent) deuterium oxide is cheaply available

The active hydrogens of —OH, —NH, and —SH groups are conveniently replaced by deuterium in the mass spectrometer itself, by repeated introduction of D_2O through the heated inlet system prior to running the sample in question. The following reaction sequences summarize some of the other more important methods of introducing deuterium:

1. $RCH_2COCH_2R \xrightarrow[\substack{\text{base} \\ \text{or phase-transfer} \\ \text{catalysis}}]{D_2O} RCD_2COCD_2R$

2. $RCD_2COCD_2R \xrightarrow[\text{Villiger}]{\text{Baeyer}} RCD_2OH + RCD_2COOH$

3. $RCH_2COOH \xrightarrow[\substack{\text{sealed tube} \\ \text{at 200°C}}]{D_2O/DCl} RCD_2COOH(D)$

4. $-CH{=}CH- \xrightarrow[D_2]{Pd/C} -CHD-CHD-$

5. $\underset{\substack{\| \\ -C-}}{\overset{O}{}} \xrightarrow[(NaBD_4)]{LiAlD_4} \underset{\substack{| \\ -CD-}}{\overset{OH}{}}$

6. $RBr \xrightarrow[(b)\,D_2O]{(a)\,Mg|Et_2O} RD$

    or $RBr \xrightarrow{Bu_3SnD} RD$

7. $R_2C{=}O \xrightarrow[(b)\,LiAlD_4]{(a)\,NH_2OH} R_2CDNH_2$

8. $RCONHR' \xrightarrow{LiAlD_4} RCD_2NHR'$

9.

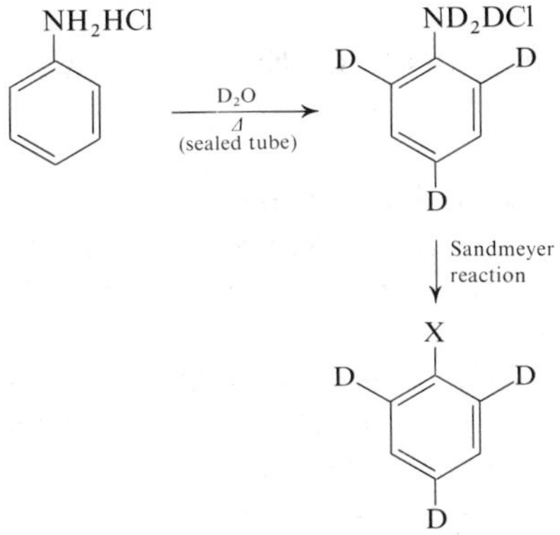

The relative ease with which specifically deuteriated analogs of organic compounds may be synthesized is illustrated by reference to the preparation of numerous deuteriated analogs of 8-methylnonan-4-one.[3]

10.

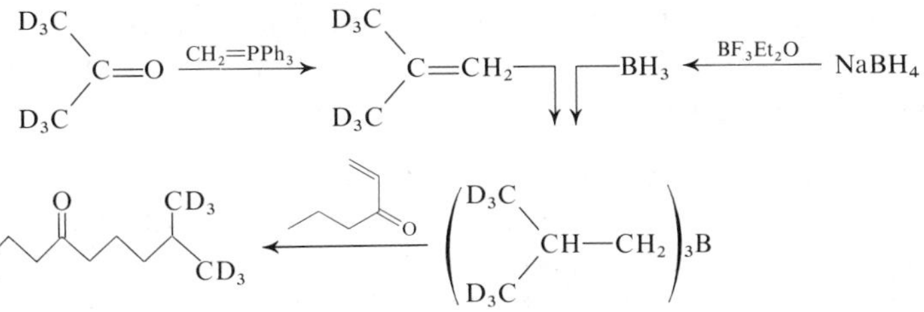

11. 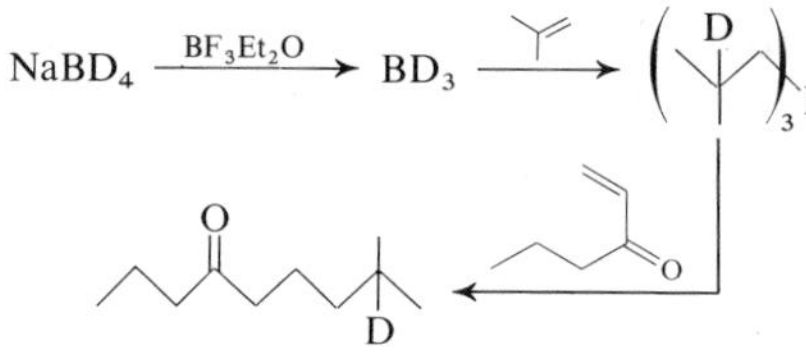 $NaBD_4 \xrightarrow{BF_3Et_2O} BD_3 \longrightarrow \left(\begin{array}{c} D \\ \end{array} \right)_3 B$

12. 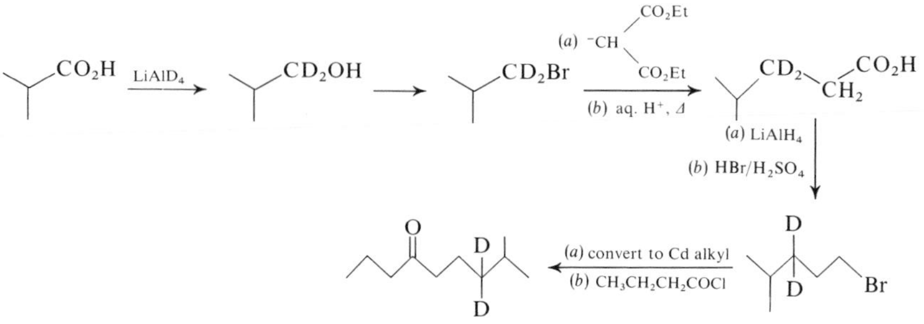 $\xrightarrow{CO_2H} \xrightarrow{LiAlD_4} \xrightarrow{CD_2OH} \longrightarrow \xrightarrow{CD_2Br} \xrightarrow[(b)\ aq.\ H^+,\ \varDelta]{(a)\ ^-CH(CO_2Et)_2} \xrightarrow{CD_2\ CO_2H\ CH_2}$

(a) LiAlH$_4$
(b) HBr/H$_2$SO$_4$

$\xleftarrow[(b)\ CH_3CH_2CH_2COCl]{(a)\ \text{convert to Cd alkyl}}$ ... Br

13. 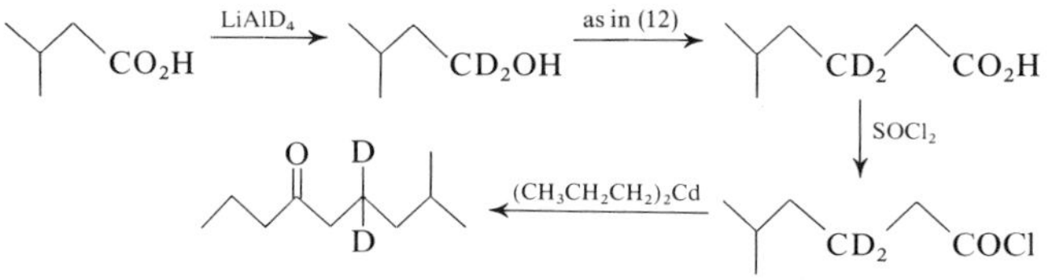 $\xrightarrow{CO_2H} \xrightarrow{LiAlD_4} \xrightarrow{CD_2OH} \xrightarrow{\text{as in (12)}} \xrightarrow{CD_2\ CO_2H}$

$\downarrow SOCl_2$

$\xleftarrow{(CH_3CH_2CH_2)_2Cd} \qquad CD_2\ COCl$

14. 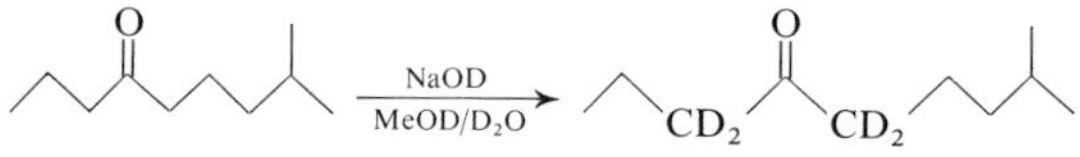 $\xrightarrow[MeOD/D_2O]{NaOD} \quad CD_2 \quad CD_2$

15. 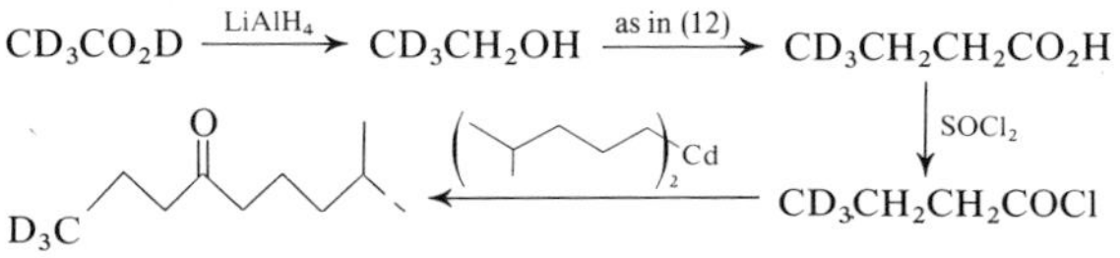 $CD_3CO_2D \xrightarrow{LiAlH_4} CD_3CH_2OH \xrightarrow{\text{as in (12)}} CD_3CH_2CH_2CO_2H$

$\downarrow SOCl_2$

$\xleftarrow{Cd} \quad CD_3CH_2CH_2COCl$

D_3C ...

B. Incorporation of ^{13}C

The cheapest source of ^{13}C for organic syntheses is $Ba^{13}CO_3$; ^{13}C-enrichment levels of 65 and 90 percent are available. Other more expensive sources are $^{13}CH_3I$ and $Na^{13}CN$ (or $K^{13}CN$).

Reaction of $Ba^{13}CO_3$ with acid liberates $^{13}CO_2$, thus providing a convenient starting material for synthetic routes to many organic compounds. In particular, carboxylic acids may be formed by carbonation of the appropriate Grignard reagent. The carboxylic acid may then be alkylated, at the α-carbon atom using 2 moles of lithium di-isopropylamine (LDA) followed by treatment with an alkyl halide; or the acid may be subjected to functional group interchange.

16. $Ba^{13}CO_3 \xrightarrow[(b)\ RMgBr]{(a)\ H^+} R^{13}CO_2H$

Specific examples of incorporation of ^{13}C into organic molecules[4] are outlined by the following schemes:

17.[4d] $Ba^{13}CO_3 \xrightarrow[\text{(b) }^{13}CH_3MgI]{\text{(a) } H^+} {}^{13}CH_3{}^{13}CO_2H \xrightarrow{(CH_3)_2SO_4} {}^{13}CH_3{}^{13}CO_2CH_3$

$$\Big\downarrow \begin{array}{c} CH_2MgBr \\ (CH_2)_3 \\ CH_2MgBr \end{array}$$

benzene ($^{13}CH_3$, ^{13}C) $\xleftarrow{Pd/C}$ cyclohexene ($^{13}CH_3$, ^{13}C) $\xleftarrow{I_2,\ C_6Cl_5OH}$ cyclohexanol ($^{13}CH_3$, ^{13}C–OH)

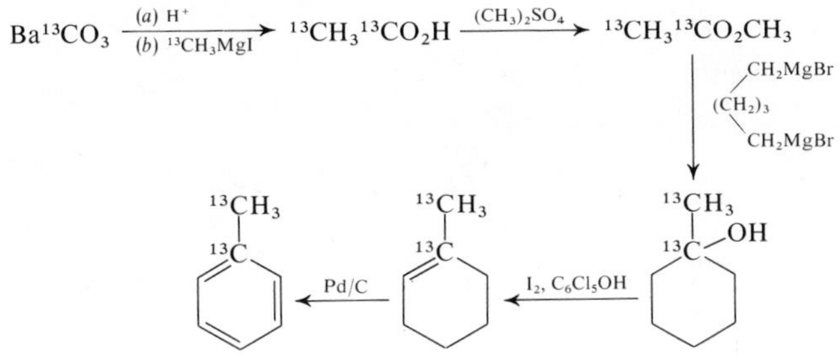

18.[4b] $Ph^{13}CHO \xrightarrow{PhCH_2MgBr} Ph\cdot{}^{13}CH{=}CH\cdot Ph$

19.[4c]

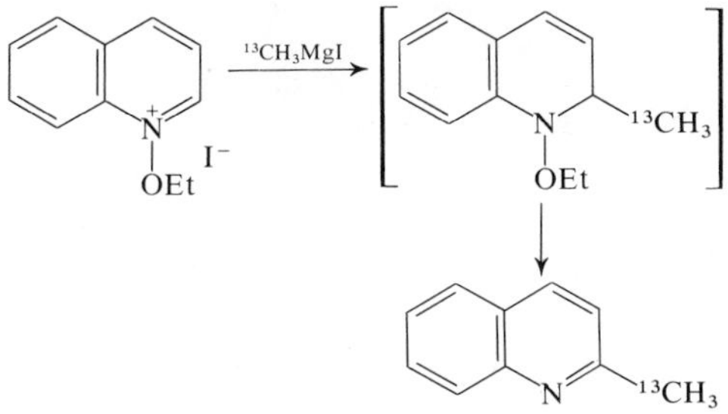

quinolinium (N$^+$–OEt) I$^-$ $\xrightarrow{{}^{13}CH_3MgI}$ [dihydroquinoline (N–OEt, $^{13}CH_3$)] $\rightarrow$ 2-methylquinoline (N, $^{13}CH_3$)

C. Incorporation of ^{18}O

Although most mass spectrometric investigations of labeled organic molecules have involved ^2H and/or ^{13}C, fragmentation mechanisms have also been studied employing ^{18}O as a label.[5] $H_2{}^{18}O$ is the normal commercial source of ^{18}O and incorporation of the label into the common oxygen-containing organic functional groups can be readily effected:

Acids and derivatives

20. $RCO_2H \xrightarrow[\text{HCl}]{H_2{}^{18}O} RC^{18}O_2H$

21. $C_6H_5CCl_3 \xrightarrow{H_2{}^{18}O} C_6H_5C^{18}O_2H \xrightarrow{SOCl_2} C_6H_5C^{18}OCl$

$$\Big\downarrow ROH$$

$^{18}O{=}C(OR)(C_6H_5)$

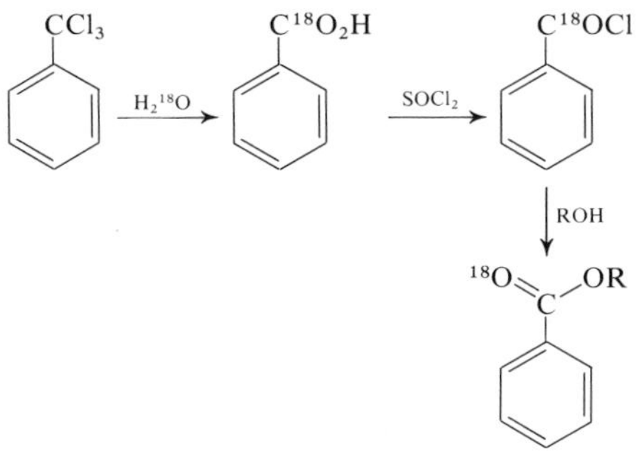

Aldehydes, ketones

22.

$$R-CO-R' \xrightarrow[\text{CF}_3\text{CO}_2\text{H}]{\text{H}_2{}^{18}\text{O/dioxan}} R-C^{18}O-R'$$

(R' = alkyl, H)

Phenols

23.

$$\text{SO}_3^-\text{Na}^+ \xrightarrow[\text{(b) H}^+]{\text{(a) Na}^{18}\text{OH}, \varDelta} {}^{18}\text{OH}$$

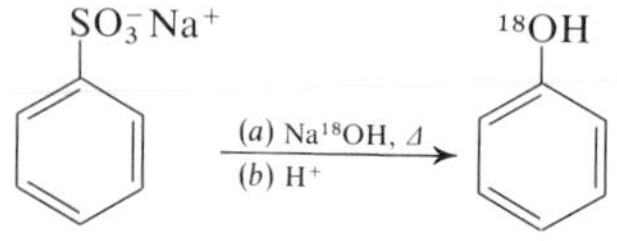

D. Incorporation of ^{15}N

Although ^{15}N has been used as a label for the elucidation of mass spectral decomposition pathways,[6] the most common use of ^{15}N in mass spectrometry is the determination of isotopic incorporation in natural products. Metabolic pathways can be deduced by introducing ^{15}N-labeled compounds (usually amino acids) *in vivo* and investigating the extent of isotopic incorporation into other metabolites.[7]

Convenient sources of ^{15}N are ^{15}NH$_3$, ^{15}NH$_4$Cl, and K^{15}NO$_3$; and amino acids labeled with ^{15}N are commercially available. A few syntheses of ^{15}N-labeled compounds are shown below:

Amides

24.

$$\text{RCOCl} \xrightarrow{{}^{15}\text{NH}_3} \text{RCO}^{15}\text{NH}_2$$

Primary amines

25.

$$\text{RCO}^{15}\text{NH}_2 \xrightarrow{\text{LiAlH}_4} \text{RCH}_2{}^{15}\text{NH}_2$$

$$\xrightarrow{\text{NaOBr}} \text{R}^{15}\text{NH}_2$$

Pyrimidines
(e.g., cytosine)

26.

$$^{15}\text{NH}_3 + (\text{C}_6\text{H}_5\text{O})_2\text{CO} \longrightarrow {}^{15}\text{NH}_2\text{CO}^{15}\text{NH}_2$$

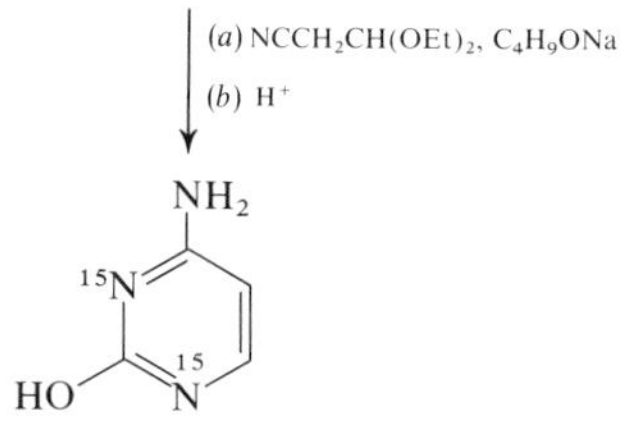

$$\Big\downarrow \begin{array}{l} (a)\,\text{NCCH}_2\text{CH(OEt)}_2,\ \text{C}_4\text{H}_9\text{ONa} \\ (b)\,\text{H}^+ \end{array}$$

11-2 REACTIONS OF ISOTOPICALLY LABELED ORGANIC IONS

A. Site-Specific Rearrangements

Deuterium labeling is a very useful technique in organic mass spectrometry for uncovering site-specific hydrogen-transfer processes. A classic example of this is found in the γ-hydrogen (McLafferty) rearrangement of ionized carbonyl compounds. Thus, the 70 eV mass spectrum of ethyl butyrate (**1**) contains a prominent ion at m/z 88, whereas for the ionized deuterium-labeled analog (**2**) this peak is shifted to m/z 89.[8] This reveals that transfer of a γ-hydrogen atom occurs (that is, **2** → **3**), but does not prove that a concerted mechanism is involved. Indeed, molecular orbital calculations suggest that transfer of the γ-hydrogen atom to oxygen may occur before the olefin molecule is eliminated.[9]

$$CH_3CH_2CH_2CO_2CH_2CH_3$$

1	**2**	**3**, m/z 89

However, it is frequently impossible to determine directly the site specificity involved in the decomposition of many organic ions. This is because interchange of atoms, bound to originally distinct sites, occurs prior to dissociation: consequently, when decomposition takes place, the atoms concerned may be randomly distributed over these various sites. As a result, even though an atom on a specific site may be transferred exclusively in the final step, this atom may not have been attached there initially. When this situation occurs, the transferred atom appears to originate from a number of sites.

For example, the McLafferty rearrangement becomes apparently nonspecific as ions of progressively longer lifetimes (e.g., metastable ions) are sampled. This could, in principle, be due to interchange of the hydrogen atoms attached to various sites before transfer to oxygen; alternatively, transfer of hydrogen atoms bound to several carbon atoms may occur from the unrearranged ion. In practice, there is much experimental evidence to support the view that interchange of atoms, initially in distinct positions, may precede dissociation. This phenomenon is common for ions of long lifetimes and is frequently encountered even for fast reactions in the ion source; its occurrence indicates that rearrangement reactions are taking place faster than the fragmentation reaction under investigation.

Nevertheless, it should not be concluded that loss of positional identity invariably precedes decomposition of metastable ions. For example, hydrogen atoms attached to heteroatoms often retain their identity completely. Moreover, hydrogen transfer through five- and six-membered ring transition states is well documented, and sometimes involves specific sites.

B. Site-Exchange Reactions

The decomposition of isotopically labeled ions occasionally exhibits no specificity whatsoever; instead, apparently random selection of the various atoms needed to constitute the eliminated neutral is observed. This behavior is especially common in hydrocarbon ions, even when aromatic nuclei are present. The most general explanation of this phenomenon is that reversible rearrangements can occur, resulting in the interchange of atoms which were originally distinct. If these isomerizations can take place rapidly, compared with the decomposition rate, all the atoms of each element become randomly distributed amongst all possible sites. Therefore, a statistical distribution is observed for the selection of some of these atoms when dissociation takes place, regardless of the structure of the reacting configuration.

For example, loss of acetylene from ionized benzene occurs with random selection of both hydrogen and carbon atoms. Thus ionized, 1,3,5-D_3-benzene (**4**) eliminates C_2H_2, C_2HD, and C_2D_2 in an approximately statistical ratio (1:3:1) at energies appropriate to metastable transitions:[10]

$$\begin{array}{ll}
\xrightarrow{-C_2H_2} C_4HD_3^{\ddagger} & m/z\ 55 \\
\xrightarrow{-C_2HD} C_4H_2D_2^{\ddagger} & m/z\ 54 \\
\xrightarrow{-C_2D_2} C_4H_3D^{\ddagger} & m/z\ 53
\end{array}$$

4, m/z 81

This result shows that rearrangement of **4** must precede decomposition; if dissociation took place from an unrearranged molecular ion, only C_2HD loss would be anticipated.

Furthermore, loss of C_2H_2, $C^{13}CH_2$, and $^{13}C_2H_2$ occurs in metastable transitions of **5** in the ratio (1:3:1) expected on the basis of statistical selection of carbon atoms.[11] These data reveal that the carbon skeleton of ionized benzene is disrupted prior to C_2H_2 loss.

5

The results found for ^2H- and ^{13}C-labeled analogs of ionized benzene could be interpreted in terms of a mechanism in which the six C—H units undergo site exchange, without cleavage of the C—H bond. However, experiments on doubly labeled ions exclude this possibility. Loss of C_2H_2, C_2HD, $C^{13}CH_2$, and $C^{13}CHD$ is observed in the ratio expected from random selection of the constituent atoms.[12,13] Hence, the C—H bonds must be broken and reformed in the course of the reversible rearrangements which lead to statistical distribution of the carbon and hydrogen atoms in ionized benzene.

The occurrence of these rearrangement processes in such a stable molecule as ionized benzene is perhaps hardly surprising. The dissociation channel with the lowest activation energy requires some $400\,kJ\,mol^{-1}$ of internal energy. The necessary rearrangement processes may well need much less internal energy, especially if simultaneous making and breaking of bonds is feasible. In fact, ring fission of ionized benzene requires less internal energy than decomposition; this conclusion stems from studies on acyclic isomers of C_6H_6, which behave in the same way as benzene in slow dissociations occurring after ionization.

In some cases, the determination of the statistical ratio of products is a simple matter (e.g., for **4** and **5**); however, in more complicated examples, the following formula is useful:

$$\frac{p!}{n!(p-n)!} \tag{11-1}$$

This formula gives the number of ways of choosing n atoms from p atoms. Thus, in the case of $C_5{}^{13}CH_5D$ (**6**), the numbers of ways of choosing C_2H_2, C_2HD, $C^{13}CH_2$, and $C^{13}CHD$ are 100, 50, 50, and 25, respectively. These figures arise from applying formula (11-1) to each of the four types of atom needed to complete the requisite neutral molecule and multiplying together the four numbers obtained. An internal check can be applied by evaluating the total number of ways of selecting two carbon and two hydrogen atoms. The value (225) must be the same as the sum $(100 + 50 + 50 + 25 = 225)$ of the ways of choosing C_2H_2, C_2HD, $C^{13}CH_2$, and $C^{13}CHD$.

6

Loss of positional integrity prior to decomposition is not confined to monocyclic aromatic molecules. For example, the molecular ion of 1-cyano-(2,4-D_2)-naphthalene (**7**) eliminates HCN and DCN in the ratio of 5 to 2 in slow reactions (see the metastable peaks at m/z 105.8 and 104.0, respectively, in Fig. 11-1). Even for fast reactions, the ratio of HCN and DCN losses is close to that expected on the basis of random selection of any one of the seven hydrogen atoms in the molecular ion.[14]

Similar behavior is observed for numerous alkyl carbonium ions. For example, the major decomposition channel of $C_4H_9{}^+$ ions, derived from any precursor, is elimination of methane at energies appropriate to metastable transitions. Loss of methane from butyl ions derived from precursors such as **8** or **9** reveals that the hydrogen and deuterium atoms are able to participate equivalently when dissociation occurs.[15,16] Similarly, ^{13}C-labeling studies show that any one of the four carbon atoms can be eliminated with equal probability from the n-$C_4H_9{}^+$ ion.[15]

$$CH_3CH_2CD_2CH_2X \qquad\qquad CD_3{-}\underset{\underset{\displaystyle CH_3}{|}}{\overset{\overset{\displaystyle CD_3}{|}}{C}}{-}X$$

8 **9**

X = Cl, Br, I, CO$_2$Me, CO$_2$Et, CO$_2$Ph   X = Br, I

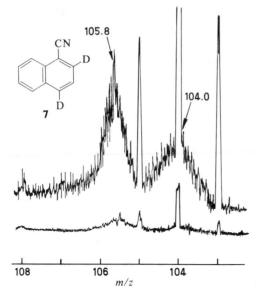

Figure 11-1

Very similar abundances are observed for the competing slow reactions of $C_4H_9^+$ ions generated from precursors having n-, iso-, sec-, or t-butyl structures. This indicates that rapid interconversion of the isomeric butyl ions precedes decomposition (see Sec. 4-2). A plausible mechanistic scheme for this interconversion is depicted below:

$$CH_3CH_2CH_2\overset{+}{C}H_2 \underset{\text{shift}}{\overset{1,2\text{-H-}}{\rightleftharpoons}} CH_3CH_2\overset{+}{C}HCH_3 \underset{\text{shift}}{\overset{1,2\text{-CH}_3\text{-}}{\rightleftharpoons}} \underset{\underset{CH_3}{|}}{\overset{+}{C}H_2CHCH_3} \underset{\text{shift}}{\overset{1,2\text{-H-}}{\rightleftharpoons}} \underset{\underset{CH_3}{|}}{CH_3\overset{+}{C}CH_3}$$

Starting from n-$C_4H_9^+$, a 1,2-hydride shift produces sec-$C_4H_9^+$; a 1,2-methyl shift then leads to iso-$C_4H_9^+$ which may rearrange to t-$C_4H_9^+$ via a further 1,2-hydride shift. The 1,2-hydride and alkyl shifts required in this mechanism are symmetry allowed;[17] furthermore, these processes are known to be extremely facile from solution NMR experiments[18] and from quantum mechanical calculations.[19] Indeed, it appears that 1,2-hydride and alkyl shifts can occur essentially without activation energy, apart from that associated with the inherent enthalpy change for isomerization. Assuming this is the case, utilization of known thermochemical data for the heats of formation of reactants and products,[20,21,22] leads to the potential energy profile shown in Fig. 11-2.[23]

It is evident from Fig. 11-2 that rapid and reversible interconversion of all four $C_4H_9^+$ isomers should occur before dissociation. Moreover, the consequences of the 1,2-hydride and methyl shifts, needed to effect these rearrangements, are that both hydrogen and carbon atoms become randomly distributed over all possible sites. Therefore, irrespective of the reacting configuration, dissociation takes place with statistical selection of the hydrogen and carbon atoms required to make up

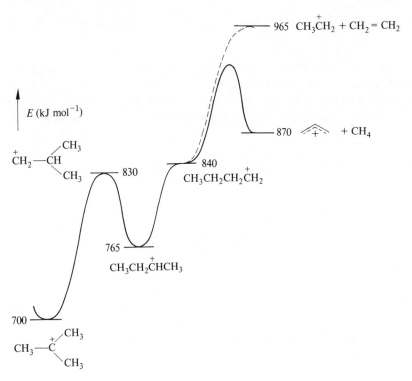

$$965 \quad CH_3\overset{+}{C}H_2 + CH_2 = CH_2$$

$E \, (\text{kJ mol}^{-1})$

$$\overset{+}{C}H_2 - \overset{CH_3}{\underset{CH_3}{CH}}$$

$$830$$

$$840 \quad CH_3CH_2CH_2\overset{+}{C}H_2$$

$$870 \quad \diagup\!\!\!\widehat{\underset{+}{\diagup}}\!\!\!\diagdown \quad + CH_4$$

$$765$$

$$CH_3CH_2\overset{+}{C}HCH_3$$

$$700$$

$$CH_3 - \overset{+}{\underset{CH_3}{C}}\!\!\diagdown^{CH_3}$$

Figure 11-2

the expelled neutral molecule. Thus, the observed statistical loss of methane from $C_4H_9{}^+$ may be understood.

The occurrence of random selection of atoms in the decomposition of organic ions is sometimes taken to mean that the ions can rearrange to a symmetrical intermediate; this conclusion is not necessarily correct. Thus, in the above example, statistical elimination of methane occurs despite the fact that no structure exists for $C_4H_9{}^+$ in which all carbon and hydrogen atoms are simultaneously equivalent. Rather, it is the rapid interconversion of isomeric ions that causes loss of positional identity.

11-3 DEUTERIUM ISOTOPE EFFECTS

A factor that must be considered in some decompositions of deuterium-labeled organic ions is the primary deuterium isotope effect. Such an effect is observed when an X—H bond is broken in the rate-determining step of a decomposition. (Most cases reported have involved C—H bond rupture, but cases are also known for X = O, N.) In view of the widespread use of deuterium labeling to determine specific hydrogen-transfer and site-exchange processes, it is important to ascertain the magnitude of the deuterium isotope effect and the conditions under which it may be expected to occur.

The origin of the primary deuterium isotope effect lies in the differences in zero-point energy between X—H and X—D bonds. On account of the higher vibrational frequency of the X—H bond, its zero-point energy is greater and less energy is required to reach the transition state for reaction. For organic ions, the effect of substituting a deuterium for a hydrogen atom is to increase the activation energy (for a C—H bond rupture) by between 0.05 and 0.14 eV. The variation of the kinetic isotope effect k_H/k_D with internal energy may be understood qualitatively by reference to the simplified rate equation:

$$k = v\left(\frac{E - E_0}{E}\right)^{s-1} \tag{11-2}$$

This leads to the isotope effect k_H/k_D in a partially deuterated organic ion:

$$\frac{k_H}{k_D} = \frac{v_H}{v_D}\left(\frac{E - E_0^H}{E - E_0^D}\right)^{s-1} \tag{11-3}$$

It is emphasized that Eq. (11-2) has severe limitations close to the threshold, but Eq. (11-3) can nevertheless give valuable qualitative information about the

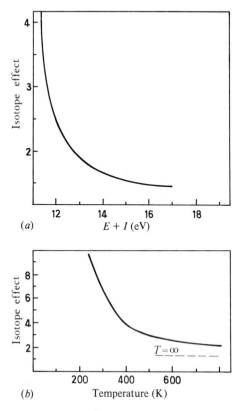

(a) $E + I$ (eV)

(b)   Temperature (K)

Figure 11-3

variation of k_H/k_D with internal energy. It is evident that (1) k_H should always be greater than k_D for the simple bond rupture, (2) k_H/k_D should decrease with increasing energy, and (3) k_H/k_D should be infinite at the activation energy for C—H rupture and should approach 1.4 at high energies ($v_H/v_D = \sqrt{2} \simeq 1.4$).

Calculations of the variation of isotope effect with internal energy have been made[24] using the improved equation of the QET; Fig. 11-3a shows a calculated curve for the loss of H and D from partially deuteriated toluenes, corrected for the number of H and D atoms in the molecular ion.[25] This variation of isotope effect with internal energy in isolated organic ions is analogous to the variation of isotope effect with temperature for thermally equilibrated organic neutrals. A calculated thermal curve for a ^{12}C—H versus ^{12}C—D bond fission is shown in Fig. 11-3b.[26]

Theory therefore predicts that high-isotope effects should be observed at ion energies just above the threshold for decomposition. This condition is satisfied for metastable decompositions, and in practice high deuterium isotope effects are often found for metastable transitions involving C—H bond cleavage. For example, CH_3D^{+}, $CH_2D_2^{+}$, and CHD_3^{+} eliminate H· exclusively in slow reactions,[27] whilst the molecular ion of CH_3CD_3 loses H· in preference to · in the ratio of 600 to 1.[28] Very large isotope effects are expected in metastable transitions from small ions, because the energy range appropriate to ions decomposing in metastable transitions is often very narrow due to the rapid rise of k with E (Secs. 3-3B and 4-4). If partial deuteriation introduces a new reaction (loss of D·, with an activation energy 0.1 eV above that for loss of H·), the higher activation energy process may require internal energies almost outside those appropriate to metastable transitions.

The evaluation of deuterium isotope effects can be complicated owing to the occurrence of only partial exchange between hydrogen and deuterium atoms. Partial exchange may take place between all possible sites; alternatively, partial or complete exchange may occur over a limited number of sites. This situation necessitates careful analysis and requires studies on species deuteriated in different parts of the molecule in question.

The mass spectra[25] of the deuteriated toluenes **10**, **11**, and **12** provide evidence for the reduction in isotope effect, with increasing internal energy, predicted by Eq. (11-3).

10	**11**	**12**

The observed ratios of H· and D· loss from the metastable molecular ions of **10**, **11**, and **12** are given in Table 11-1. These ratios must then be corrected by

Table 11-1 $[M^+ - H]/[M^+ - D]$ **for unimolecular metastable ions from deuteriated toluenes at 70 eV**

	10	11	12
$[m_H^*]/[m_D^*]$ (observed)	4.68 ± 0.09	1.68 ± 0.09	4.77 ± 0.15
Observed/statistical	2.81 ± 0.06	2.80 ± 0.15	2.86 ± 0.09

the appropriate statistical factor to allow for the different numbers of hydrogen and deuterium atoms in the ions. These factors are the ratios of losses which would be expected from ionized **10**, **11**, and **12** on the basis of statistical participation of hydrogen and deuterium atoms and no isotope effect (that is, 5/3 for **10** and **12**, 3/5 for **11**). No assumption whatsoever need be made about the structure of the reacting configuration for decomposition.

The corrected ratios (observed/statistical) are the same within experimental error. These data are consistent with: (1) reversible interconversion of the molecular ion of toluene with other structures, thus resulting in complete H/D exchange over all the original sites, prior to decomposition; (2) equal participation of H· and D· loss, with an isotope effect of 2.8 favoring H· elimination. Similar conclusions stem from an independent study,[29] where a somewhat larger isotope effect of 3.5 was deduced. The slightly larger isotope effect found in the second study presumably arises because the population of ions dissociating in metastable transitions has a longer lifetime; consequently, the ions have a lower average internal energy and a larger isotope effect is observed. It is notable that the isotope effect is not very large, in contrast to those found for ionized methane and ethane (see above). This is because the rise of k with E is slower for ionized toluene, which has more degrees of freedom than ionized methane or ethane. Therefore, the range of internal energies of metastable toluene molecular ions is greater than those appropriate for the metastable molecular ions of methane and ethane.

The relative abundances of $M^+ - H·$ and $M^+ - D·$ ions, formed in fast reactions of ionized **10**, **11**, and **12** occurring in the ion source, are given in Table 11-2. The corrected ratios (observed/statistical) reveal that: (1) the hydrogen and deuterium atoms do not become equivalent prior to dissociation—instead, there is a preference for eliminating a hydrogen atom from the original methyl group (see also Fig. 11-4); and (2) the isotope effect is smaller than that operating in the decomposition of metastable ions.

Table 11-2 $[M^+ - H]/[M^+ - D]$ **for daughter ions from deuteriated toluenes at 70 eV**

	10	11	12
$[M^+ - H]/[M^+ - D]$	1.76 ± 0.06	1.14 ± 0.02	2.90 ± 0.04
Observed/statistical	1.06 ± 0.04	1.90 ± 0.03	1.74 ± 0.02

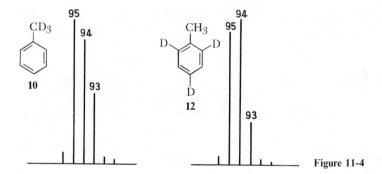

Figure 11-4

11-4 BIOLOGICAL APPLICATIONS OF ISOTOPE LABELING

The use of mass spectra of isotopically labeled species in elucidating the mechanisms of biosynthetic pathways is elegantly illustrated by the following example.[30]

The biosynthesis of prostaglandin E_1 (**14**) from 8,11,14-eicosatrienoic acid (**13**) in the presence of atmospheric oxygen involves incorporation of three atoms of oxygen. To investigate the origin of the oxygens at carbon atoms 9 and 11, [2-^{14}C]8,11,14-eicosatrienoic acid was incubated with a preparation of vesicular gland for 10 min in an atmosphere containing $^{18}O_2$. The trihydroxy derivative of **14** was then prepared by adding ice-cold enthanolic $NaBH_4$ to the incubation mixture. After esterification (ethyl ester) and methylation of the hydroxy groups, followed by permanganate–periodate oxidation and further esterification, the diester **15** was formed:

The mass spectra of a reference sample of **15** and of a sample of **15** formed in an atmosphere containing $^{18}O_2$ were then compared. The spectra were obtained using a directly linked GC/MS system (Sec. 8-2), about 5 µg of material being injected into the column. Three fragment ions which retained both oxygen atoms in the ring were evident: at m/z 343 $[M-CH_3]^+$, 313 $[M-OC_2H_5]^+$, and 297 $[M-(CH_3+C_2H_5OH)]^+$. In the spectrum of **15** derived from prostaglandin E_1 formed in an atmosphere containing $^{16}O_2$ (43 percent), $^{16}O^{18}O$ (1 percent), and $^{18}O_2$ (56 per cent), peaks were also seen at m/z 347, 317, and 301. Furthermore, these peaks (due to fragments containing two ^{18}O atoms) were 12 to 17 times more abundant than the peaks at m/z 345, 315, and 299, clearly demonstrating that the oxygen atoms of the hydroxyl group at carbon atom 11 and of the keto group at carbon atom 9 originate from the same molecule of oxygen.

Other biological applications include the use of labeled reagents to deduce the numbers of sites in a molecule which may be methylated or acetylated. For instance, separate samples can be derivatized using unlabeled and labeled diazomethane, acetic anhydride, and methyl iodide in conjunction with an appropriate base, or combination of these reagents. Comparison of the two spectra obtained may yield valuable information concerning the nature of the compound of interest.[31]

REFERENCES

1. H. Budzikiewicz, C. Djerassi, and D. H. Williams: "Structure Elucidation of Natural Products by Mass Spectrometer," vol. 1, Holden-Day, San Francisco, 1964; for an account of deuterium labeling, see A. F. Thomas, "Deuterium Labelling in Organic Chemistry", Appleton-Century-Crofts, New York, 1971.
2. A. Murray and D. L. Williams: "Organic Syntheses with Isotopes," Interscience, New York, 1958.
3. G. Eadon and C. Djerassi: *J. Am. Chem. Soc.*, vol. 92, p. 3084, 1970.
4. See, for example: (a) R. G. Cooks and S. L. Bernasek: *J. Am. Chem. Soc.*, vol. 92, p. 2129, 1970; (b) P. F. Donaghue, P. Y. White, J. H. Bowie, B. D. Roney, and H. J. Rodda: *Org. Mass Spec.*, vol. 2, p. 1061, 1969; (c) P. M. Draper and D. B. Maclean: *Can. J. Chem.*, vol. 48, p. 746, 1970; (d) K. L. Rinehart, A. C. Buchholz, G. E. Van Lear, and H. L. Cantrill: *J. Am. Chem. Soc.*, vol. 90, p. 2983, 1968.
5. See, for example: (a) R. H. Shapiro and K. B. Tomer: *Org. Mass Spec.*, vol. 3, p. 333, 1970; (b) J. H. Bowie and P. Y. White: *J. Chem. Soc.*, vol. B, p. 89, 1969; (c) T. H. Kinstle, O. L. Chapman, and M. Sung: *J. Am. Chem. Soc.*, vol. 90, p. 1227, 1968.
6. See, for example: (a) R. T. M. Fraser and N. C. Paul: *J. Chem. Soc.*, vol. B, p. 1407, 1968; (b) P. N. Rylander, S. Meyerson, E. L. Eliel, and J. D. McCollum: *J. Am. Chem. Soc.*, vol. 85, p. 2723, 1963.
7. See, for example: (a) A. K. Bose, K. G. Das, P. T. Funke, I. Kugajevsky, O. P. Shukla, K. S. Khanchandani, and R. J. Suhadolnik: *J. Am. Chem. Soc.*, vol. 90, p. 1038, 1968; (b) P. D. Shaw and J. A. McCloskey: *Biochemistry*, vol. 6, p. 2247, 1967.
8. K. Biemann: "Mass Spectrometry: Organic Chemical Applications," p. 121, McGraw-Hill, New York, 1962.
9. F. P. Boer, T. W. Shannon, and F. W. McLafferty: *J. Am. Chem. Soc.*, vol. 90, p. 7239, 1968.
10. K. R. Jennings: *Zeitschrift für Naturforschg*, vol. 22a, p. 454, 1967.
11. I. Horman, A. N. H. Yeo, and D. H. Williams: *J. Am. Chem. Soc.*, vol. 92, p. 2131, 1970.
12. R. Dickinson and D. H. Williams: *J. Chem. Soc.*, vol. B, p. 249, 1971.
13. W. O. Perry, J. H. Beynon, W. E. Baitinger, J. W. Amy, R. M. Caprioli, R. N. Renaud, L. C. Leitch, and S. Meyerson: *J. Am. Chem. Soc.*, vol. 92, p. 7238, 1970.
14. R. G. Cooks, I. Howe, S. W. Tam, and D. H. Williams: *J. Am. Chem. Soc.*, vol. 90, p. 4064, 1968.
15. B. Davis, D. H. Williams, and A. N. H. Yeo: *J. Chem. Soc.*, vol. B, p. 81, 1970.
16. A. N. H. Yeo and D. H. Williams: *Chem. Comm.*, p. 737, 1970.
17. R. B. Woodward and R. Hoffmann: *Angew. Chem. Int. Edn.*, vol. 8, p. 781, 1969.

18. G. A. Olah and P. v. R. Schleyer (eds.): "Carbonium Ions," Wiley and Sons, New York; see especially chap. 33 (vol. 4).
19. See, for example: L. Radom, J. A. Pople, V. Buss, and P. v. R. Schleyer: *J. Am. Chem. Soc.*, vol. 94, p. 311, 1972.
20. F. P. Lossing and G. P. Semeluk · *Can. J. Chem.*, vol. 48, p. 955, 1970.
21. F. P. Lossing: *Can. J. Chem.*, vol. 49, p. 357, 1971.
22. J. L. Franklin, J. G. Dillard, H. M. Rosenstock, J. T. Herron, K. Draxl, and F. H. Field: "Ionization Potentials, Appearance Potentials and Heats of Formation of Gaseous Positive Ions," National Bureau of Standards, Washington, D.C., 1969, app. I; see also: H. M. Rosenstock, K. Draxl, B. W. Steiner, and J. T. Herron: *J. Phys. Chem. Ref. Data*, vol. 6, suppl. 1, 1977.
23. D. H. Williams: *Accounts Chem. Res.*, vol. 10, p. 280, 1977.
24. M. Vestal and G. Lerner: "Fundamental Studies Relating to the Radiation Chemistry of Small Organic Molecules," Aerospace Research Laboratory Report 67-0114, U.S. Defence Documentation Center, Alexandria, Va., 1967.
25. I. Howe and F. W. McLafferty: *J. Am. Chem. Soc.*, vol. 93, p. 99, 1971.
26. H. Eyring and F. W. Cagle: *J. Phys. Chem.*, vol. 56, p. 889, 1952.
27. L. P. Hills, M. L. Vestal, and J. H. Futrell: *J. Chem. Phys.*, vol. 54, p. 3834, 1971.
28. U. Löhle and Ch. Ottinger: *J. Chem. Phys.*, vol. 51, p. 3097, 1969.
29. J. H. Beynon, J. E. Corn, W. E. Baitinger, R. M. Caprioli, and R. A. Benkeser: *Org. Mass Spectrom.*, vol. 3, p. 1371, 1970.
30. B. Samuelsson: *J. Am. Chem. Soc.*, vol. 87, p. 3011, 1965.
31. E. Hunt and H. R. Morris: *Biochem. J.*, vol. 135, p. 833, 1973.

CHAPTER

TWELVE

CHOOSING A MASS SPECTROMETER SYSTEM

The early chapters of this book have discussed various instrumental configurations and have developed the theory surrounding the formation, fragmentation, and analysis of organic ions in the mass spectrometer. These chapters were followed by a presentation of the relationship between fragmentation pattern and structure for the different organic functional classes. In later chapters, a critical discussion was given of those peripheral instrumental techniques (particularly GC and computers) that have enabled mass spectrometry to encompass such a wide field of analytical applications in recent years.

No two sets of applications are identical and the appropriate mass spectrometer system must be chosen with care. This final chapter builds upon the information provided in previous chapters and attempts to aid the user of mass spectrometry in choosing a suitable system. No reference is made to specific manufacturers or their products.

12-1 DEFINING THE APPLICATIONS

A. Type of Sample

Since different types of mass spectrometer encompass widely differing *mass ranges*, it is important to estimate the masses of samples likely to be encountered. These will range from below mass 100, where gases or volatile liquids are being analyzed, up to several thousand mass units for some natural products. The *volatility* of the sample is partly related to its mass but should be considered as a separate factor. This is because high sample polarity may indicate the use of a "soft" ionization technique but not always necessitate a high-mass range. The volatility of the sample also influences the choice of inlet system.

A crucial parameter governing the choice is the *purity* of the sample, since mixtures usually require specialized inlet systems (for example, GC or LC interfaces). A related factor is the time for which the sample is available for obtaining a mass spectrum. Individual mixture components may, in some cases, require scan periods as low as 1 s. Another consideration is the amount of sample available and the consequent *sensitivity* required.

B. Information Required

In the early days of organic mass spectrometry, a full mass spectrum of the compound under study was invariably obtained. This was usually necessary in order to identify an unknown compound. Nowadays, however, the mass spectrometer is frequently used as a sensitive detector for compounds that are suspected to be present (see Sec. 8-2C). Under these conditions only selected ions are monitored and the user should establish whether the *full spectrum* or *selected ion monitoring* is required. On the occasions where full spectra of known compounds are obtained it might be considered advantageous to compare these against computer-stored *libraries* of spectra.

The *mass resolution* and *accuracy of mass measurement* required should also be carefully considered. A distinction between these two parameters is made here since it is becoming increasingly possible to determine accurate masses (for elemental composition determinations) from microprocessor-controlled scans on low-resolution instruments. However, if separation of peaks of the same integral mass is required then a *high-resolution* mass spectrometer is essential. On the occasions where metastable-ion spectra (for example, MIKE spectra) are to be obtained, then it should be established what *energy resolution* is required.

The user should determine whether the main requirement is for *molecular weight information* on the particular samples investigated. The choice of ionization method rests partly on this decision.

Perhaps the fastest-growing application of mass spectrometry nowadays is in *quantitative measurements*, particularly in the life sciences. It is important to estimate the accuracy of the quantitative measurements required for the amounts of sample likely to be encountered.

C. Physicochemical Applications

The applications of many mass spectrometer systems are directed solely to analytical problems. However, a large number of mass spectrometry groups carry out research into fundamental processes of ion chemistry occurring in the mass spectrometer (see also Chaps. 3 to 5). This research is carried out either on a full-time basis or in conjunction with service work, and sometimes requires specialist apparatus.

Perhaps the commonest physicochemical applications in mass spectrometry are different kinds of energy measurements. The *kinetic energy release* in metastable transitions can be determined from metastable peak widths and *ionization energies* are measurable in many mass spectrometers. Thermodynamical quantities, notably proton affinities and basicities, are also obtainable from measurements on ion-molecule reactions.

This section has only touched on the many physical measurements which can be made with the mass spectrometer. As with all applications, the potential purchaser should ascertain that the instrument yields the required accuracy.

D. Future Applications

The most difficult factor in the choice of a mass spectrometer system is the prediction of future trends and applications. Technology in the field is constantly improving, particularly where computers are involved. Consequently, every effort should be made to determine (1) the future requirements of the user and (2) the future capabilities of mass spectrometer systems in fulfilling these requirements. Another important consideration in this respect is the most probable lifetime of the mass spectrometer system purchased.

12-2 TYPE OF INSTRUMENTATION

When the main applications of the proposed mass spectrometry system have been defined, the next step is to identify the instrumental units appropriate to these applications.

A. Inlet Systems

The inlet systems (usually more than one per instrument) chosen for a particular range of applications are determined mainly by two factors: (1) the sample volatility and (2) whether the sample is pure or a mixture. When volatile mixtures (often derivatized) are analyzed, a GC/MS interface (see Sec. 8-2) is desirable and where the mixture components are more polar, LC/MS inlet systems are becoming acceptable (see Sec. 8-3). Unless analysis is confined entirely to mixtures, provision of a direct insertion probe for involatile samples of medium/high purity and/or a glass, heated inlet system for volatile, pure samples is desirable. The provision of a spare probe is advisable as a safeguard against breakages. Where a data system is employed it is, of course, necessary to incorporate some form of inlet system for the volatile calibrant.

An excessive number of different inlet systems should be avoided on one instrument. The added complexity, particularly where chemical ionization is employed and therefore a gas-tight seal between the ion chamber and the inlet lines is required, tends to result in more instrument down-time. Multiple-inlet lines on magnetic instruments operating at high accelerating voltage may also promote high-voltage breakdown (sparking) in the ion source.

B. Ion Sources

Electron-impact (EI) and *chemical ionization* (CI) ion sources are usually of comparable sensitivity and the choice between them rests mainly on whether it is essential to generate abundant ions in the molecular ion region (see Secs. 2-2, 7-1, and 7-2). If single-ion monitoring and/or quantitative measurements are being carried out, then there is a strong case for employing CI. For certain compound classes, e.g., underivatized sugars, CI yields much more useful spectra than EI. Another occasion for utilizing a CI source is the quantitative investigation of ion-molecule reactions at high pressures.

Most commercial CI sources operate also in the EI mode but such ion sources are not designed specifically to operate at high performance in the EI mode without a CI gas, although the situation is changing. Consequently, it is preferable to purchase two separate sources (one EI, one dual EI/CI) where economic circumstances allow. Although CI is now used routinely, two drawbacks are as follows:

1. Source lifetime may be low, particularly where hydrocarbon reagent gases are employed.
2. Source discharges may occur with magnetic instruments operating at high accelerating voltage.

Field desorption sources are not generally as sensitive as EI or CI sources and are not as consistently reliable. Nevertheless, for polar compounds an FD spectrum may yield indispensable molecular weight information. A useful accessory to an FD source is a kit for welding and growing activated FD emitters (which are expensive when purchased direct from the mass spectrometer companies). Since activation of tungsten emitters usually takes about eight hours and is not always successful, it is advisable to purchase (or construct) an emitter preparation unit that accommodates several emitters simultaneously.

Two other types of source are suitable for certain applications. Special EI sources for operation at *low electron beam energies* ($< 15\,$eV) are marketed for analysis of compounds (e.g., aromatic hydrocarbons) having low ionization energies and abundant molecular ions. Another variation is the adaptation of a CI source for *negative-ion* analysis (see Sec. 7-2).

A few general points should be considered with all types of source. Some modern ion sources are fitted with viewing windows or transparent plates to facilitate fault finding. Other advantages are accessibility and ease of cleaning. Finally, spare sources are a valuable asset to reduce down-time during source cleaning and should be purchased if financial circumstances allow.

C. Analyzers

For most applications, the choice lies between a magnetic or quadrupole mass analyzer. If low-mass (< 800) analyses at low resolution (< 1000 RP) form a high percentage of the samples to be run, then there is a strong case for purchasing a quadrupole analyzer. Quadrupoles tend to be somewhat more reliable than magnetic analyzers and do not require high accelerating voltages. For those numerous GC/MS applications where low-mass, low-resolution spectra with fast scan times are required, quadrupole analyzers are sold in large numbers. However, some magnetic instruments are capable of matching the performance of quadrupoles for such applications.

Where high-resolution, high-mass spectra are required a double-focusing magnetic instrument is essential. Spectra obtained from such instruments in good condition are of far better quality than those obtained from quadrupole instruments. The price of such magnetic instruments is accordingly higher. However, if ultra-high mass resolution (40,000–150,000 RP) is not vital, the needs of the mass spectroscopist can often be met from the medium-resolution (5,000–40,000 RP) range of double-focusing magnetic instruments. The methods for determining accurate masses by employing a data system were discussed in Sec. 9-2F, but it should be emphasized that the highest accuracy is currently attainable via a manual peak-matching accessory. Therefore, it is always preferable to include this device in a double-focusing magnetic instrumental package. The double-beam mass spectrometer (see Sec. 7-3) is a medium-resolution instrument and its unique features make it an attractive proposition within its price range. With appropriate data system software it is capable of accurate mass measurements (15 ppm at 3,000 RP) at mass resolutions lower than those applicable on single-beam instruments. Its double-source feature also permits the capability of low-resolution mass counting for FD, CI, and negative-ion spectra, which usually contain few peaks. The difficulty in counting low-resolution mass spectra, particularly at high mass, without an appropriate data system or mass marker is sometimes overlooked when purchasing an instrument. In determining an unknown structure, it is imperative in the first instance to be able to identify m/z values correctly to the nearest integer.

Double-focusing magnetic mass spectrometers may be of so-called conventional or reversed geometry. The features of these respective instruments have been discussed in Sec. 3-2 and 8-4. Various metastable scan facilities are available with both types of instrument and the merits of such accessories, together with appropriate collision cells, should be carefully evaluated before purchase.

A vital accessory for highly sensitive detection techniques and quantitative measurements is a multiple-ion-detection system. In such a system (available with both magnetic and quadrupole analyzers, under microprocessor control if required), the number of peaks monitored and the speed of switching between them should be high enough to fulfill the needs of the user.

The only other type of mass analyzer in common use is found in the ion-cyclotron resonance (ICR) mass spectrometer (see Sec. 1-3D). The ICR spectrometer is suitable for the study of ion-molecule reactions and their thermodynamics at relatively low pressures (compared with pressures found in CI courses). The mass resolution of such instruments is usually low, although Fourier transform ICR promises to alleviate this problem.

D. Data System

There are certain applications in mass spectrometry for which a data system (see Chap. 9) is almost essential. The most important of these are GC/MS or LC/MS analyses where it is possible to accumulate vast amounts of data in a short period of time. On these occasions, a data system is an enormous time saver and is therefore almost indispensable. It is important that the computer in such a system possesses sufficient storage capacity.

Another occasion where data systems are of crucial importance is in library searching. Where a small library ($<1,000$ spectra) is searched continually with almost certain success (e.g., in the perfume industry or in drug-overdose identification), then a data system with a library search facility is a valuable time saver.

The facilities available with a mass spectrometry data system are numerous (see Chap. 9 and particularly Table 9-2) and the number of options purchased with one particular system are often governed by economic circumstances. However, both software and hardware options are frequently additive and may be purchased in several stages.

E. Vacuum System

An efficient vacuum system is essential for the effective operation of the mass spectrometer. This is most commonly fulfilled by use of a combination of rotary and oil diffusion pumps, although on some modern mass spectrometers turbo-molecular pumps provide high performance despite their simplicity.

Satisfactory answers to the following questions should be established before purchase:

1. Is the pumping capacity sufficient for operation of GC, LC, and CI facilities?
2. Is there an adequate protection system against vacuum accidents?
3. Is the baking system efficient and easy to operate?
4. Is there a sufficient number of ion gauges to monitor the pressure throughout the instrument?
5. Are there any weak points in the vacuum system design that are likely to result in loss of vacuum?

F. Multiple Instrumentation

The foregoing discussions in this chapter have shown that far more mass spectrometric facilities exist than can be accommodated in one instrument. If funds allow, a multiple mass spectrometer system (consisting of two or more mass spectrometers with an appropriate data system) can reduce this problem. In such arrangements, each mass spectrometer is usually dedicated to different applications. A useful combination would be, for example,

1. a high-resolution, double-focusing magnetic instrument with EI (and CI or FD) sources and a probe inlet and
2. a quadrupole spectrometer with EI and CI sources fitted with a GC interface.

Both spectrometers would be coupled to one data system with full real-time facilities. Multiple systems incorporating more than two mass spectrometers should further improve efficiency, since not only could each instrument be dedicated to a particular type of problem but also instrument down-time could be absorbed more easily.

G. Future Additions and Developments

When a new mass spectrometer system is being considered, it is prudent to ascertain whether the new system will accommodate any future improvements and developments. The following questions are among those currently applicable:

1. Will the proposed ion source construction accommodate developments in LC technology?
2. Will the present GC interface cope with improvements in GC technology?
3. Is the present pumping capacity sufficient to cope with CI if it should be added later; if not, can the additional capacity be simply fitted?
4. If the mass range of a magnetic instrument is to be extended by development of a larger magnet, will it be interchangeable with the current magnet?
5. Is the modification for analysis of negative ions simple if required?
6. Will it be readily possible to place the mass spectrometer under micro-processor control at a future date?
7. Can the present data system be additively updated should more mass spectrometers be coupled to it, without a large portion of system software and hardware becoming redundant?

The above list is by no means exhaustive and it is intended merely as a guide to the type of questions to be considered.

H. Checklist

The foregoing sections in this chapter have defined some of the applications and sample requirements in mass spectrometry and have related these to the types of instrumentation available. This information has been condensed in tabular form and is shown as a checklist in Table 12-1.

12-3 MAKING THE CHOICE

When the best *type* of instrumentation has been defined, it is then time to compile a detailed dossier on the appropriate mass spectrometer systems offered by the

Table 12-1 Mass spectrometer check list

Sample types and requirements	Appropriate instrumentation																
	GC/MS	LC/MS	Direct probe	AGHIS	CI	FD	Magnetic analyzer	Double focusing	Peak matching	Reversed geometry	Collisional activation	Double beam	Quadrupole analyzer	Data system	Library search	Microprocessor control	Multiple ion detection
Mass > 1000							●●										
Mass < 1000							●						●				
Pure solids			●●														
Pure liquids				●●													
Volatile mixtures	●●													●●			
Involatile mixtures		●								●	●			●●			
Sensitive detection					●											●	●●
Known compounds															●		
Polar samples		●			●	●											
Molecular weights					●	●											
High-mass resolution				●●			●●	●	●				●	●			
Accurate mass							●	●				●	●	●		●	
"Metastable" studies							●●			●	●						

Key: ● suggested consideration, ●● almost essential.
Electron-impact source is assumed.

different instrument companies. This is advisable, not only for the buyer's own benefit but also for consideration by the committee that may confirm or fund the application. A weak or ill-considered case may be rejected.

A. Sources of Information

The three main sources of information for evaluating the relative merits of different instruments are: (1) sales literature, (2) results from the instrument companies' applications laboratories, and (3) users' experience.

1. A careful perusal of the sales literature will give the buyer a good indication of the potential performance of the system to be purchased. Numerous performance parameters (e.g., sensitivity, mass resolution, mass range, scan speeds, capacity of data system) are specified, and these specifications must of course be met on installation. However, evaluation of the sales literature and discussions with salesmen are only a beginning and it is strongly advisable to follow procedures (2) and (3) detailed below.
2. It is good practice to submit typical samples from the buyer's laboratory to the various instrument companies, to be run in their applications laboratory. The potential customer should endeavor to be present when the samples are run in order to see at first hand any advantages or drawbacks.

   Since all mass spectrometer systems are subject to fluctuations in day-to-day performance, disappointment in results from the applications laboratory on one particular day must not be allowed to cloud the overall evaluation. However, failure to run the samples adequately within a few weeks should be taken as a serious detrimental factor.
3. The most important sources of information in contrasting different instruments are existing users, since they are able to provide a reliable reflection of instrument performance. As many users as possible should be consulted for all the instruments being considered and the instrument companies themselves are usually able to provide lists of such operators.

B. Miscellaneous Considerations

Economic factors are often uppermost in the mind of the purchaser. Sometimes terms may be negotiated with the instrument companies which involve, for example, purchase of peripheral equipment at reduced rates or part exchange of existing equipment.

The customer should maintain a healthy wariness of *new lines* of equipment. Since such equipment is unproven in the field, the buyer ought to be satisfied that the company is sound and that the new instrument at least retains some reliable features of existing instrumentation. It might be considered prudent to wait until several new instruments are in the field and initial problems have been eliminated. However, it should be borne in mind that better financial terms may occasionally be obtained with new lines of mass spectrometer equipment.

The quality of *after-sales service* must be investigated. Spare parts and expert service should be quickly obtainable. In addition, the relative expense of parts and service ought to be evaluated for the different companies. It is sometimes possible to obtain these at reduced rates from other than the manufacturer of the mass spectrometer.

It is essential that the *contract* of purchase provides adequate safeguards. Sufficient warranty (usually one year) should be provided to cover the malfunctions which inevitably occur with a new mass spectrometer system. Some manufacturers may be tardy in delivering the instrument and it may be advantageous to write clauses into the contract that penalize late delivery.

The mass spectrometer must be given a suitable *location* at the purchasing institution. Magnetic instruments, in particular, are heavy and their performance is subject to external vibrations. For these reasons ground-floor or basement locations are often the best. In addition, magnetic fields may influence one another and magnetic instruments should not be placed too close together. When a potential site has been found, the instrument company would usually wish to confirm its suitability in any event.

The adequate *maintenance* of a mass spectrometer system requires a considerable backup of appropriate staff. Apart from personnel to operate the equipment, it is an asset to have an electronics expert and a glassblower available. Some users rely heavily on their own expertise, but where service from the manufacturer is utilized a financial provision should be made on purchase. An appropriate service contract is sometimes advisable.

The expense of *consumable items* should not be overlooked when budgeting for purchase of the mass spectrometer system. Expenditure is incurred not only on hardware incorporated into the spectrometer itself (e.g., pump oil, filaments, gold rings) but also on consumable items peripheral to the equipment (e.g., gases for CI, GC). Air-conditioning is also desirable to maintain the equipment at constant temperature and humidity.

Finally, the successful operation of a mass spectrometer system depends only partly on the after-sales service provided by the instrument company. The user should treat the mass spectrometer with care, following all the set procedures for running samples and routine maintenance. It is advantageous to maintain contact with other users in the field, both informally and by attendance at users' meetings organized by the manufacturers.

INDEX